D0378939

The Perfect Wave

Päs, H. (Heinrich)
The perfect wave : with
neutrinos at the bounda
2014.
33305230384772
cu   04/22/14

The Perfect Wave

With Neutrinos at the Boundary of Space and Time

HEINRICH PÄS

Harvard University Press

Cambridge, Massachusetts
London, England
2014

Copyright © 2014 by the President and Fellows of Harvard College
All rights reserved
Printed in the United States of America

An earlier version of this work was first published as
Die perfekte Welle: Mit Neutrinos an die Grenzen von Raum und Zeit,
copyright © 2011 by Piper Verlag GmbH, München.

Library of Congress Cataloging-in-Publication Data

Päs, H. (Heinrich)
The perfect wave : with neutrinos at the boundary of
space and time / Heinrich Päs.
       pages   cm
Includes bibliographical references and index.
ISBN 978-0-674-72501-0 (alk. paper)
1. Particles (Nuclear physics)—History.   2. Neutrinos—Mass.
3. Cosmology.   4. Space and time.   I. Title.
QC793.16.P37   2014
539.7'215—dc23   2013025630

To my father, who inspired in me the curiosity for
the things beyond the horizon, and the love of nature.

To my mother, who inspired in me the
love of beauty—and of books.

Contents

Preface ~ ix

1 Dawn Patrol in Honolulu ~ 1

2 Eleusis, Plato, Magic Mushrooms ~ 7

3 Quantum Physics: The Multiverse of Parmenides ~ 17

4 Black Dots on a White Background:
   The Particle World ~ 34

5 Beyond the Desert: Symmetries and Unification ~ 51

6 From Symmetry Breaking to Supersymmetry ~ 65

7 Birth of an Outlaw: The Neutrino ~ 81

8 Nuclear Decays a Thousand Meters Underground ~ 97

9 New Physics Is Falling from the Skies ~ 115

10 Cosmic Connections ~ 131

11 Neutrinos: Key to the Universe ~ 154

12 Extra Dimensions, Strings, and Branes ~ 167

13 Einstein's Heritage: What Is Time? ~ 188

14 How to Build a Time Machine ~ 202

15 Against Hawking and the Timekeepers ~ 213

16 Into the Wilderness of the Terascale ~ 230

17 Epilogue: Major Tom and the Singing Socrates ~ 248

Notes ~ 257

Further Reading ~ 265

Acknowledgments ~ 279

Index ~ 283

Preface

Warning up front! This book deals with established scientific insights and with wild speculations. There is motivation, even justification, for the speculations, but they are neither proven nor confirmed by experiment. While much of the established science itself can sound crazy when first encountered, I highly recommend that the reader be alert to this difference.

Well, all I can say is that at times truth is stranger than fiction.
—Vikas Swarup, *Slumdog Millionaire*

Of the topics treated within this book, quantum physics in Chapter 3, particle physics in Chapter 4, and the importance of symmetries in Chapters 5 and 6 reflect the recent scientific state of knowledge, as does the discovery of neutrinos and their masses, treated in Chapters 7 and 9. This is also true for the theories of special and general relativity in Chapter 13 and the major part of cosmology described in Chapter 10. The promotion of the concept of symmetries to GUT theories and supersymmetry as well as cosmic inflation is speculation which nevertheless represents the present hopes and expectations of most of the scientists working actively in these fields. String theory and extra dimensions are speculations, too, considered to be "wilder" by many physicists since they deviate more strongly from established theories or are more difficult to probe experimentally. Shortcuts in extra dimensions are speculations squared, and time travel in extra dimensions is a speculation . . . well, at least to the power of three, but many established physicists consider it rather as speculation to the power of 1,000. On the other hand, physics has always been full of surprises and apparent paradoxes: Simultaneity depends on the observer, rigid causality no longer exists, and clocks run more slowly when one is in motion. It is not at all impossible that the physics of time travel will be just as baffling.

Scientific speculations are great fun, but they are more than that. They are an integral part of science itself; they guarantee that science can actually progress. Naturally, such speculations require a critical mind that examines carefully what one dares to believe, what one wants to analyze more accurately, and what one is ready to discard as unlikely. But if one wants to capture science in its actual liveliness and fascination, one cannot refrain from giving speculations at least a little bit of coverage.

The wild ideas of yesterday quickly become today's dogma.
—SHELDON GLASHOW,
Winner of the Nobel Prize in Physics, 1979

Like Proust be an old teahead of time.
—Jack Kerouac, "Belief and Technique for Modern Prose"

Dawn Patrol in Honolulu

As a postdoc you have the craziest job in the world.

You get your PhD with honors at one of the best universities, and at the same time they kick you out. You apply everywhere in the world, and you have not the faintest clue on which continent or in which time zone you will end up. Then you get a new job at a different university, and you can come and go whenever you want and do whatever you want. You can just think about the most esoteric questions that come to your mind: What is all matter made of? What is happening at the beginning and at the end of the universe? What is time? All that is required is that you write down whatever insights you gain and publish them in scientific journals. Until two years later and irrespective of how good you were—then you get kicked out again, and the whole story repeats itself.

This was the reason I was standing, on this February morning's dawn, on a beach on the other side of the world. With my Danish friend Ketil I stared, shivering, into Waikiki's surf, rolling in from the endless Pacific (Fig. 1.1). Finally one

Figure 1.1. View over Waikiki to Diamond Head crater. (Courtesy Ullstein Bild/Prisma/Steve Vidler)

of us brought himself to push his board into the water and started to paddle out toward the break, while the sun slowly climbed over the rim of Diamond Head crater.

Surfing in Hawai'i ranks somewhere between national sport and religion—an active ceremony honoring the Hawaiian god Lono.[1] What is most impressive for me about surfing, though, is the suddenness of the experience. In the beginning the water is often totally glassy, but after paddling out for a few hundred meters you reach the reef. Bathed in the first sunlight, a tiny fluctuation created by a storm thousands of miles away grows—as if caused by magic—into a wave. Soon the wave transmutes into an impressive greenish blue wall, a wall that—if after many failed tries, with tired arms and a headache, you manage to catch—accelerates you breathlessly and throws you into an ocean of adrenaline as it breaks into a carpet of snow-colored whitewater.

Physics ideas are like the Hawaiian surf. They seem to come from nowhere, they catch you suddenly without warning, and they draw you into an exhilarating trip. Later, on such days, I would sit in seminars daydreaming of breaking waves, or I would wander restlessly over the Manoa campus of the University of Hawai'i, gazing at the cliffs overgrown by rain forest, while my brain untangled a clutter of film clips of the surf and thoughts about my present physics project. On one of these afternoons the following thought occurred to me: Assume that our universe, with its three space and one time dimension is embedded in a larger space with many extra dimensions. Our universe, lying there like the carpet in your living room. Could this carpet buckle? Assume further that there are elementary particles that can propagate not only in our common 3+1 dimensions but in the whole extra-dimensional space. Could these particles—instead of following the carpet buckles—take a shortcut along the floor? Could they make a "duck dive," like a surfer diving underneath an incoming wave while paddling out? Could they in this way be faster than everything moving along the bumpy surface? How would it look, then, for an observer on the carpet, or the ocean-surface universe, if these particles traveled even faster than light? Could they actually—according to a well-known paradox in Einstein's relativity—travel back in time?

At first glance, any single part of this idea seems crazy. Why should the universe have more than three space dimensions? And even if there were reasons, where in the world should these extra dimensions be? What's the nature of these exotic elementary particles? And why should our universe have even the slightest similarity to a buckled carpet or the surface of the Pacific Ocean on a winter day? Why should something that moves faster than light be able to travel backward in time? How could this be possible at all?

This is all part of a long story, whose beginnings date back more than 3,000 years to ancient Greece. A story that continued in the 1920s at Europe's universities, and has been augmented by science fiction and recent experiments at the most advanced laboratories for elementary-particle physics—CERN in Geneva, Switzerland, and Fermilab near Chicago. A true story that at times resembles a mystery thriller, with secret cults, dead philosophers, mind-expanding drugs, the mysterious disappearance of brilliant scientists, subterranean research labs, and experiments in Antarctica. It is a story of an elementary particle that, just like the Silver Surfer in the superhero cartoons, surfs to the boundaries of knowledge, of the universe, and of time itself. A story that captivates you as it sucks you into a maelstrom like an oversized wonderland.

Jump on your board and hold tight.

2

Eleusis, Plato, Magic Mushrooms

Three-thousand-year rewind: It was the ancient Greeks who began to consider the world as an object and who undertook its first rational analysis. It was they who laid the foundation for the scientific method and the Western worldview. It seems, however, that the Greeks also suffered from a divorce of ego and world, of subject and object. Deeply hidden in their souls, it seems, was the need to perpetuate some portal to the ecstatic feel of the unity of everything that exists, of all being. As the German philosopher Friedrich Nietzsche enthused in the nineteenth century: "Did those centuries when the Greek body flourished and the Greek soul foamed over with health perhaps know endemic ecstasies? Visions and hallucinations shared by entire communities or assemblies at a cult? How now? . . . Should it have been madness . . . that brought the greatest blessings upon Greece?"[1]

Ecstasies, visions, and hallucinations had a place of their own in ancient Greece: About thirty kilometers down a highway northwest of Athens, situated between crude-oil refineries and a military airfield, the small city of Elefsina hosts the remains of what was probably the most secret and holy cult, then called Eleusis, of the ancient world. "Happy are they among earthly mortals who have seen these things, but who is uninitiate and who has no part in them never has such joys once dead, beneath in darkness and gloom,"[2] says a Homeric hymn.

For more than 2,000 years Eleusis initiated up to 3,000 believers annually, among them Roman emperors and politicians such as Cicero, Hadrian, and Marcus Aurelius; poets such as Pindar, Pausanias, and Sophocles; philosophers such as Plato; but also slaves, prostitutes, and normal citizens. The rite celebrated the cycle of life and of the seasons, the unity of life and death.[3] While the death penalty was imposed on those uncovering its secrets, its traces can be found even today in science and the arts.

According to legend, Persephone, daughter of Demeter, the goddess of fertility and harvest, had been raped by Hades, the god of death, while picking flowers. With the permission of Zeus, Hades abducted her into the underworld. On earth, meanwhile, her mournful mother Demeter let the fields wither, the flowers fade, and god and man starve. Finally Zeus, the lord of Mount Olympus, gave in and commanded his brother Hades to bring Persephone back to her mother and to the world of the living. The sly Hades obeys the command but feeds Persephone a pomegranate seed on her way up to the surface. When joyful Demeter at their reunion learns that Persephone ingested food in the realm of the dead, she realizes that her daughter cannot stay with her permanently: Everyone who eats food in the realm of the dead falls under the spell of the powers of death. Eventually Zeus and Demeter agree on a compromise: One-third of the year Persephone has to stay with Hades in the underworld, while Demeter mourns and winter governs earth, but two-thirds of the year Persephone can stay with her mother, during which time the fields bear plenty of fruit.

The celebration of the cyclic recurrence of Persephone evolved from its origins in archaic rites of fertility into an allegory of death and birth and the cycle of life. On top of that, it also marks the transition from the worship of Mother Earth, personified by Demeter, to the worship of a full panoply of Olympian gods and goddesses, which illustrates the crossover of an agricultural society into a society of warriors, leading eventually to modern individuation. And finally it symbolizes—through the reunion of the two goddesses—the recovery of the original unity within the cycle of nature, long lost in daily life even in those ancient times, thus loosening the tension that seems deeply implanted in the soul of ancient Greece.

For the aspirants in Eleusis, the cultic celebrations belonged among the most important experiences in their entire lives. In early summer they had already attended the minor mysteries and recognized Persephone's death, preparing themselves for the major festivities in the late summer. They had fasted, washed themselves in the ocean, sacrificed a young pig, and, adorned with myrtle crowns, had followed the procession of the cult objects and statues of the gods on the holy street. On top of the bridge crossing the small river they had been abused and ridiculed. After their arrival in Eleusis they were consecrated to the goddess and followed the different stations of Demeter's quest: horror, sorrow, and the joy of being reunited. They danced at the fountain and drank the kykeon, a brew of barley and mint; were once more purified with fire and air; were allowed to touch the holy womb and phallus; and witnessed the dreadful appearances at the entrance of Hades's grotto, the portal to the underworld. Finally, after some time of searching, they found the sanctification hall, the Telesterion (Fig. 2.1), where they took the oath next to the holy fire and witnessed the holy showing of the cult objects. The highlight of the holy night was the appearance of Persephone herself. After the sacrifice of a male sheep, the death goddess was called up and eventually emerged in

Figure 2.1. View into the sanctification hall, the Telesterion in Eleusis.

her incarnation as the goddess of fertility, illuminated in brilliant light, and carrying with her her son Brimos, the seed of life, back from the underworld up into the light. According to the cyclic symbolism of the ritual act, Demeter is identical to Persephone, and Brimos is Hades is Dionysus.[4] The god of grain and prosperity is identical to the god of wine, fertility, and ecstasy. The father is equal to the son and thus the *archetype of indestructible life.*[5]

The comprehension of the Eleusinian cult, and in particular of the mysterious kykeon brew, took an unexpected twist when, on April 19, 1943, a young Swiss chemist started on what is now a legendary bicycle ride.[6] Albert Hofmann was a rather conservative fellow, a person who would send a jar of homemade honey to his favorite poet, Ernst Jünger. Professionally he had experimented with different derivatives of lysergic acid, the common component of several alkaloids found in a cereal fungus known as ergot. When synthesizing lysergic acid diethylamide—in short, LSD—Hofmann noticed that his laboratory animals acted nervously, and he himself felt funny during a second synthesis a few years later. To explore this effect thoroughly, Hofmann took the risk of a self-experiment with the smallest dose for which he expected any reaction at all. What he didn't know, though, was that LSD has an effect about a hundred times stronger than comparable psychoactive substances, and in this way he stoned himself into nirvana: Within less than an hour on his bicycle trip home from the lab, the serotonin neurons stopped firing while the postsynaptic neurons inside the locus coeruleus acknowledged every stimulus, with wild fusillades bombarding the limbic system and consequently the entire brain: "If the doors of perception were cleansed, every thing would appear to man as it is, infinite," wrote William Blake in *The Marriage of Heaven and Hell*.[7] "Break on through to the other side!" sang the Doors.[8]

My only personal experience with magic mushrooms started with a tram that would not operate since on its route a house was burning down and ended with my friend Olaf massaging the feet of a girl of nodding acquaintance on my sofa while I sat for half an hour in the bathroom laughing about the wall tiles. What I found so exhilarating was that what under normal circumstances would be a rather boring rectangular pattern became blurred and kept changing shape, transforming over and over into new arrangements and increasingly crazy and creative designs. On top of that, green dots were flickering on the white walls like emeralds, the outlines of all objects obtained an unaccustomed clarity, colors seemed to glow from inside, and the heater warped itself into the aggressive form of a futuristic weapon.

Persons who, like Hofmann, take LSD, mescaline, or magic mushrooms in high doses often give accounts of the following sensations:

- an extremely enhanced perception, in which objects wander in and out of the foreground depending on the focus of concentration;
- extreme colors and patterns made out of numerous repetitions of the same elements or geometrical shapes;
- synesthesia (the capability to see sounds as colors or shapes or to hear physical contact); and

- decomposition of the ego and distortions of the awareness of space and time.

A friend once told me how, after trying *Salvia divinorum,* he merged with the floor and spent an eternity living the life of a carpet. The British author Aldous Huxley described such altered states of consciousness as follows: "In the final stage of egolessness there is an 'obscure knowledge' that All is in all—that All is actually each."[9]

And it was Hofmann's first "trip" that turned his life upside down. Hofmann was convinced of the usefulness of LSD, and before long there was a coming and going of potheads, artists, hippies, and crackpots in his home—among them the drug apostle Timothy Leary, who had just escaped from his prison in California. Later, Hofmann's experience with LSD inspired him to isolate and analyze the active ingredients of the Indian cultic mushroom *teonanácatl* and the "magic" seeds *ololiuqui.* He even set out on a horse expedition into the Mexican highlands in order to search for the magic plant *Salvia divinorum.*[10] And eventually he teamed up with the ethnomycologist R. Gordon Wasson and the classical scholar Carl Ruck to work out a revolutionary and controversial interpretation of the incidents that occurred in the Telesterion of Eleusis: The mind-altering experiences reported by the aspirants could have originated from ergot, the fungus from which the essential constituent of LSD had been recovered.[11]

The three scholars had good reasons for this idea. First, it seemed rather likely that some hallucinogenic drug was involved, since thousands of believers experienced a holy show on one specific night. Next, a scandal in the fifth century BCE had been uncovered, in which the sacrilegious consumption of kykeon took place in private, profane festivities, suggesting both an important role for this brew in the course of the ritual and a psychoactive component qualifying it as a party drug. A third hint for the possible significance of ergot came from the importance of cereals in the cult of Demeter and the symbolic interpretation of the infected grain as sexual intercourse. Further hints for the sacred significance of mushrooms and fungi can be found in various ancient texts: The Greeks named mushrooms food of the gods; Hades was pictured as a god with violet hair, reminiscent of the violet stipe of the full-grown ergot mushroom; and the name of one of Demeter's husbands, Iasion, can be translated as "man of the drug." Hofmann also pointed out that the hallucinogenic lysergic acid alkaloids of ergot were soluble and could easily be extracted in water, while the related poisonous components were not. And finally, symptoms of nausea, dizziness, trembling limbs, and cold sweat, as well as the nature of the vision experienced by the aspirants in Eleusis, pointed toward a psychedelic drug such as LSD or magic mushrooms.

If one contemplates how many generations of the elite of the ancient world lived through such experiences—an elite that considerably shaped the origins of thought and science of the entire Western world—it is not too surprising that traces of these incidents can be found in the testimonies and scriptures of the Greeks. For example, Greek tragedy has been understood as an allegory for the cosmic cycle of life and death.[12] The chorus symbolizes the eternal essence of the universe, a Dionysian commotion or world music. This chaotic state is being resolved temporarily into the beautiful appearance of the story of the protagonists who, however, eventually have to die again, becoming unified with the Dionysian entity of the universe. For Nietzsche this idea became a symbol for the course of life: Life had meaning only as a vision of a universal chorus, justified only by the aesthetic quality of its narrative. In a similar way, Hofmann speculated that the Greeks overcame their suffering about the divorce of ego and world and resolved it in the mystical holistic experience induced typically by psychedelic drugs, which weaken or even annihilate the boundary between subject and object.[13] In fact, many concepts advocated in ancient Greek philosophy seem to mirror the influence of psychedelic drugs—above all, the teachings of Plato, initiate of Eleusis and probably the most influential of all Greek philosophers. In his famous "Allegory of the Cave," he compared people with cave dwellers who perceive only the shadows of their environment and have to climb up into the light to perceive the true nature of the world—something that has been dubbed Platonic "mystery lingo."[14] The mystic holistic experience leads to the idea of the unity of all that is, to the assumption that the entirety is more than the sum of its parts, and eventually to the possibility that this One, which remains unaffected and unchanged behind the fluctuating reality, can be comprehended in a mystical experience. Indeed, such ideas constitute a central theme in Greek philosophy, starting with their origins in the work of Parmenides and Heraclitus in the pre-Socratic era, their adoption by Plato, their reappearance in the Neoplatonism of the late ancient world, and their influence on all of Western metaphysics. Moreover, the perception of smallest details during a drug trip may have affected the atomism of Democritus, who first assumed that the world should be built up of indivisible elements. Also, the reduction of the world around us into elementary patterns and symmetries can be found as the central theme of Plato's *Timaeus*. Even a cyclic consciousness of time and history has been ascribed to the Greeks of the classical and pre-classical eras, and has been celebrated by Nietzsche as "eternal recurrence," in crude contrast to the modern belief in eternal advancement.[15]

Already in the cradle of Western culture the Greeks were time-travelling—at least according to their own worldview. It is a moot question whether a preexisting philosophy and attitude toward life influenced the Eleusinian ecstasies or, on the contrary, the ancient Greek worldview was essentially inspired by the mind-expanding experience. In any case, the mutual relation of Greek soul and Eleusinian ecstasy makes for an intriguing topic. And, even more intriguing—as incredible as it may sound—these thoughts had a crucial bearing on the outlook of a small circle of physicists who, more than twenty centuries later, laid the foundation for the modern understanding of the microcosm with what seemed like the craziest theory ever heard of: quantum mechanics.

3

Quantum Physics
The Multiverse of Parmenides

A major breakthrough in the story of quantum physics begins with a young man holed up in a rain pipe—in order to find a quiet place for reading. It is the year 1919, in Munich, shortly after the end of World War I. The chaotic rioting in the streets that followed the revolution driving the German emperor out of office has finally calmed down, and now eighteen-year-old Werner Heisenberg can find some leisure time again.[1] He had been working as a local guide, assisting a vigilante group that was trying to reestablish order in the city, but now he could retreat, after the night watch on the command center's hotline, onto the roof of the old seminary where his cohort was accommodated. There he would lie, in the warm morning sun, in the rain pipe, reading Plato's dialogues. And on one of these mornings, while Ludwig street below him and the university building across the way with the small fountain in front slowly came to life, he came across that part in *Timaeus* where Plato philosophizes about the smallest constituents of matter, and the idea that the smallest

particles can finally be resolved into mathematical structures
and shapes, that one would encounter symmetries as the basic pillar of nature—an idea that will fascinate him so deeply
that it will capture him for the rest of his life.[2]

Werner Heisenberg (Fig. 3.1) was to become one of the
most important physicists of his generation. When just turned
forty, he was the head of the German nuclear research program, which in World War II examined the possibilities for
utilizing nuclear power, including the feasibility of nuclear
weapons. In this position he was on the assassination list of
the US Office of Strategic Services, but a special agent who
had permission to kill Heisenberg in a lecture hall decided
against it, after he heard Heisenberg's lecture on abstract S-ma-
trix theory and concluded that the practical usefulness of
Heisenberg's research was marginal.[3] Even today, historians
debate Heisenberg's role in Nazi Germany. His opponents
criticize his remaining in Germany and his commitment to
the nuclear research project, the so-called *Uranverein,* which,
according to these critics, failed to build a nuclear weapon for
Hitler only because Heisenberg was unable to do it.[4] Extreme
admirers, such as Thomas Powers in *Heisenberg's War,*[5] argue
that Heisenberg used his position to prevent the construction
of a German nuclear bomb by exaggerating its difficulties
when questioned by officials, bestowing a moral mantle on
Heisenberg he never had claimed for himself.

What is well documented is that Heisenberg traveled to
Copenhagen, Denmark, in the fall of 1941 to visit his fatherly
friend and mentor Niels Bohr. According to Heisenberg, his
intention was to inform Bohr that the construction of a nuclear
bomb was possible but that the German physicists would not

Figure 3.1. From left to right: Enrico Fermi, godfather of the neutrino; Werner Heisenberg, a creator of quantum mechanics; and Wolfgang Pauli, the father of the neutrino. (Photo by Franco Rasetti, courtesy AIP Emilio Segre Visual Archives, Segre Collection, Fermi Film Collection)

try to build it and to suggest that physicists in the allied nations should follow the same policy.[6] This epic conversation, however, only resulted in a lasting breakdown of their friendship. Bohr, the son of a Jewish mother and the citizen of an occupied country, could not have much sympathy for any agreement with the German physicist.

In 1998, the British author Michael Frayn wove different perceptions of this meeting into a play that essentially deals with the parallel existence of different realities, both in psychology and in quantum mechanics.[7] After all, among all his other activities, Heisenberg was famous for one thing: He was one of the masterminds of a revolutionary new theory.

Just six years after the sunny morning in the rain pipe, Heisenberg, now twenty-three years old and a postdoc at the University of Göttingen, was forced by his hay fever to leave his institute for two weeks, and he spent some sleepless time on Helgoland, a tiny and once holy red rock off Germany's coast in the North Sea—days that would shatter the most basic grounds of physics. One-third of the day the young man climbed in the famous cliffs; one-third he memorized the works of Goethe, the poet who served as a national idol in Germany and who followed the classical paradigm of the ancient Greeks; and the last third he worked on his calculations. In these calculations he developed a formalism that would be the bedrock of modern quantum physics and would do nothing less than change the world: "In Helgoland there was one moment when it came to me just as a revelation. . . . It was rather late at night. I had finished this tedious calculation and at the end it came out correct. Then I climbed a rock, saw the sun rise and was happy."[8]

Nowadays the technical applications of quantum physics account for about one-third of the US gross domestic product. Nevertheless, Richard P. Feynman commented some forty years after Heisenberg's work that the theory is so crazy that nobody can actually comprehend it,[9] and Einstein had earlier declared bluntly: this is obvious nonsense.[10] What makes quantum physics special is that this theory breaks radically with the concept of causality. In our daily lives we are used to ordered sequences of cause and effect: You and a friend clink your glasses with just a little bit too much verve; one glass breaks; beer runs down to the floor; your significant other/roommate/parents cry out. One event causes the next one. This is exactly where quantum physics is different, where this strict connection between cause and effect no longer exists. For example, how a particle reacts to an influence can be predicted only in terms of probabilities. But this is not the end of the story: Unless the effect on the particle is actually observed, all possible consequences seem to be realized simultaneously. Particles can reside in two different locations at once! And particles exhibit properties of waves while waves behave in certain situations like particles. An object thus has both properties of a particle and of a wave, depending on how it is observed. The particle corresponds to an indivisible energy portion of the wave, a so-called *quantum*. On the other hand, the wave describes the probability for the particle to be located at a certain place. This property of quantum mechanics can be depicted most easily with the famous double-slit experiment (Fig. 3.2).

When a particle beam hits a thin wall with two narrow slits in it, the corresponding wave penetrates both slits and spreads out on the other side as a circular wave. On a screen situated behind the wall, in accordance with the wave nature of the electrons, an interference pattern appears, resulting from the superposition of the waves originating from the two

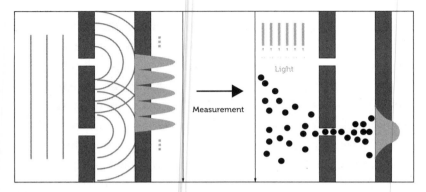

Figure 3.2. Double-slit experiment. As long as no measurement determines which slits the particles are passing through, they behave like interfering waves, which pass simultaneously though both slits (left side). Where two wave crests coincide, the probability of detecting a particle is largest; where a crest coincides with a trough, the probability is very small or zero. The resulting image is called an interference pattern. As soon as an external measurement disturbs the system—for example, if one uses irradiation with light to determine which path the electrons take through the slits—the wave collapses into single particles, which accumulate in narrow bands behind the slits they were flying through (right side).

slits in the wall. Where a crest meets another crest or a trough meets another trough the wave gets amplified. A crest encountering a trough, on the other hand, results in little or no amplitude (left side). This pattern appears, however, only as long as it is unknown through which slit a single electron passed. As soon as this is determined, for example by blocking one of the slits or by irradiating the electrons with light, the two-slit interference pattern gets destroyed and the electrons behave just like classical particles. To be more accurate, a new wave emanates from the slit, and the pattern exhibited on the screen is the one for a wave passing through a single slit, which resembles a smooth probability distribution (right side).

Heisenberg and Bohr interpreted this as a collapse of the wave function due to the measurement process in which one gets a result with the probability given by the amplitude squared of the wave. This is the so-called *Copenhagen interpretation of quantum physics*, which is still taught at universities around the globe. According to this interpretation, a particle is located in many places simultaneously until finally a measurement assigns it a concrete location. And this is true not only for position; it applies to other measurable quantities such as momentum, energy, the instant of a nuclear decay, and other properties as well. Erwin Schrödinger, both collaborator with and competitor of Heisenberg in the development of quantum physics, carried this idea to an extreme: "One can even set up quite ridiculous cases. A cat is penned up in a steel chamber, along with the following device (which must be secured against direct interference by the cat)." In Schrödinger's experiment the death or life of the cat depends on whether a radioactive isotope does or doesn't decay in a particular time period. As long as we do not check whether the isotope did decay or not, nor how the cat is doing, Schrödinger's cat is simultaneously dead and alive, or as Schrödinger phrased it: "[The wave function of the system would have] in it the living and dead cat (pardon the expression) mixed or smeared out in equal parts."[11]

There are two reasons why we don't observe such bizarre phenomena in our daily lives: One is that the wavelengths of ordinary objects around us are tiny compared with the sizes of the objects themselves. The other is that the objects we deal with every day are always interacting with their environment and thus are being measured all the time. A beer bottle, for example, may very well be situated in two different locations, but only for an extremely short time and for an extremely small separation (too short and too small to measure).

Bohr summarized the apparent paradox of particles and waves under the concept of complementarity. After a guest lecture he gave at Moscow University, he left the following aphorism on the blackboard where famous visitors were supposed to leave comments: "Contraria non contradictoria sed complementa sunt" (Opposites do not contradict but rather complement each other).[12]

But back to Heisenberg, Plato, and the ancient Greeks: As the American philosopher of science Thomas S. Kuhn realized, science in times of scientific revolutions is particularly vulnerable to nonscientific influences.[13] When changes to the scientific paradigm cause a shift in the generally accepted problems and solutions and thus also in the general perception and scientific world view, rational reasons like conformity with facts, consistency, scope, simplicity, and usefulness are not sufficient to understand the evolution of a new theory. During these times, personal factors such as cultural background can also play a decisive role. And Heisenberg's background was almost as Greek as it was German: As the son of a professor of Greek language, he became accustomed to Greek philosophy and culture and their reception in early twentieth-century Germany long before he himself learned Latin and ancient Greek in school. His biographer Armin Hermann suggests that the encounter with Plato's philosophy influenced Heisenberg more than anything else.[14] And not long after Heisenberg studied, climbed, and calculated in Helgoland, Paul Dirac in Cambridge and Erwin Schrödinger in Vienna worked out different but mathematically equivalent versions of quantum physics. Since the standard interpretation of these works was developed basically in the inner circle around Bohr and Heisenberg, Heisenberg's background seems particularly relevant for its appreciation. Also, Schrödinger made statements such as "Almost our entire intellectual heritage is of Greek origin" and "science can be correctly characterized as reflecting on the Universe in a Greek way."[15] And Dirac left on the blackboard in Moscow, right next to Bohr's principle of complementarity, only the laconic remark, "A physical law has to have mathematical beauty,"[16] a statement that reminds us strongly of Goethe's transfiguration of the classical worldview:

Nature and art, they seem each other to repel
Yet, they fly together before one is aware;
The antagonism has departed me as well,
And now both of these seem to me equally fair.[17]

And sure enough, quantum physics seems to be a Greek theory after all. This becomes evident when reading the thoughts in the book *Die Einheit der Natur (The Unity of Nature)* by Heisenberg's student and friend, Carl Friedrich von Weizsäcker,[18] the brother of the subsequent German president, on the centerpiece of quantum physics—the wave-particle duality—and how it can be traced back to the arguments in Plato's dialogue *Parmenides.*

Parmenides of Elea (Fig. 3.3) was a Greek philosopher in the pre-Socratic era around the fifth century BCE. Of his writing only the fragment of a philosophical poem remains;[19] it deals with the unity of all being. It describes how an unnamed

Figure 3.3. Parmenides of Elea (center) in Raphael's *The School of Athens.*

goddess—often understood as Persephone—invites the poet to perceive the truthful being—again a likely reference to the mystical experience in the mystery cults of Eleusis. The truthful being then is distinguished from mere appearances and described as the all-embracing One—uncreated and indestructible, alone, complete, immovable, and without an end[20]—reminiscent of Aldous Huxley's stage of egolessness.[21] "Ἕν τό πᾶν" (One is the All) is correspondingly the central statement followed up by Weizsäcker[22] when he discusses the argument between Socrates and Parmenides chronicled by Plato,[23] which, according to the Italian author Luciano De Crescenzo, was the "most boring and complicated discussion in the entire history of philosophy."[24]

In this battle of words, which supposedly took place on the occasion of a visit of Parmenides to Athens, Socrates tried to refute the identity of One and All. To this end Socrates argued that One is not Many and thus has no parts. On the other hand All refers to something which does not miss any of its parts. Consequently the One would consist of parts if it were All, and thus finally One cannot be the All.[25]

At this point Weizsäcker comes to Parmenides's defense by stressing the connection with quantum mechanics. And it is really astounding how the quantum mechanical interpretation of the One suddenly bestows this incomprehensible debate with lucidity and meaning. After all, in quantum mechanics the All is the wave function and, in its fullest manifestation, the all-embracing wave function of the universe.

Moreover, in quantum mechanics the analysis of the individual parts of an object without destroying the object is impossible, since the measurement, as explained above, affects the object and thus distorts the unity of its parts. And of all possible states an object can assume, only an infinitesimally small fraction are states in which the parts of the object actually correspond to a clearly defined outcome of a measurement. Only in these states can one truthfully assign reality or existence to these parts. For example, only two among the infinitely many possible states that Schrödinger's cat can assume (such as 90 percent alive and 10 percent dead or 27.3 percent alive and 72.7 percent dead)—namely totally dead or totally alive—correspond to possible outcomes in a measurement. But quantum mechanically, a pair of two cats, half of them dead and the other half alive, is realizable not only with one living and one dead but also with two half-dead cats or one being 70 percent alive and one being 30 percent alive. Consequently, in quantum physics the All is really more than its parts, the partial objects actually constituting through their association a new entity, or, just as postulated by Parmenides, a new unity, a new One. Now Parmenides, according to Plato, required further that the One possesses no properties: It has no beginning, no center and no end, no shape and no location; it is neither in itself nor in anything else; it is neither at rest nor is it moving. Weizsäcker can argue that a quantum mechanical object fulfills these requirements perfectly. After all, a determination of any of these properties relies on a measurement, which implies a collapse of the wave function and thus destroys the unity of the collective object. On the other hand, isolation of the object from the surrounding universe is impossible:[26] The object would not exist in the universe if it were not connected to the universe via some kind of interaction.

Thus, strictly speaking, only the universe as a whole can constitute a real quantum mechanical object. Then, however, nobody would remain who could observe it from outside. Next Weizsäcker and Parmenides follow the discussion backward: how the One—meaning the all-embracing universe barring all properties—unfurls into the colorful and multifaceted appearances of our everyday life. The argument relies here on the quirky assumption that the One, in the instant where it "is"—in the sense of exists—is already two things. It is the One and it is the Is. This argument can be iterated. Again both the One and the Is are two things: the Is is and is the One, and the One is and is the One. By repetition of this consideration the One acquires an infinite multiplicity: The being One unfolds itself into the universe. And again Weizsäcker clarifies the discourse by referring to the quantum mechanical object. After all, the way an object can exist is via interaction with other objects, which again results in the collapse of the wave function and the loss of quantum mechanical unity: In order to establish that an object exists, the object has to be measured and thus is affected in a way that implies that it is no longer one object according to the meaning of Parmenides's One. In summary, Weizsäcker arrives at an amazing conclusion, that the notion of complementarity has its source in ancient Greece: "We find . . . the foundation of complementarity already foretold in Plato's *Parmenides.*"[27] We actually can recover the feel of what the ancient Greeks experienced in their mystery cults in modern twentieth-century physics!

But this is not the end of the story: The atomism of Democritus, the idea that the world is not continuously divisible but made out of indivisible particles, makes sense only in the context of quantum mechanics, where matter consists of compound objects that correspond to standing waves and thus can absorb or emit energy only in indivisible portions— the quanta. Also, the idea of tracing the laws of nature back to fundamental symmetries, as proposed in Plato's *Timaeus,* is, as we will see in Chapter 5, an integral part of contemporary particle physics. Finally, consider Einstein's objection to the fundamental importance of probabilities. Because of that objection, he remained a lifelong opponent of quantum mechanics: God doesn't play dice with the world.[28] This statement appears as a direct response to the 2,500-year-old fragment of Heraclitus: "Eternity is a child moving counters in a game; the kingly power is a child's."[29]

How can one really comprehend the lack of causality inherent in quantum physics and in particular the role of the puzzling quantum collapse, which are not described by the mathematical formalism and remain controversial today? The most modern and consistent interpretation of these puzzling phenomena seems to be at the same time the craziest one: Everything that can happen does happen—albeit in different parallel universes. This idea was formulated for the first time in 1957 by Hugh Everett III while he was working on his doctoral dissertation at Princeton University. With the bizarre concept of parallel universes he asked too much of his contemporary physicists, even though Everett—like Richard Feynman, a founder of quantum electrodynamics, and Kip Thorne, the father of the wormhole time machine—was a student of the eminent John Archibald Wheeler, who was himself a rather unorthodox and creative associate of Einstein and who, among many other achievements, coined the term *black hole* for the timeless star corpses in the universe. But even with this first-class mentor, Everett's colleagues didn't take him seriously. Everett left the academic world shortly after finishing his dissertation. During a frustrating visit in Copenhagen, during which Everett tried to convince Niels Bohr to take some interest in his work, he (Everett) transformed a standard approach in classical mechanics into a method for optimization that he could apply to commercial and military problems and that helped him to become a multimillionaire—but didn't make him happy. He became a chain-smoking alcoholic and died of a heart attack when he was only fifty-one years old. According to his explicit wish, his ashes were disposed of in the garbage. Fourteen years after his death, his daughter Elizabeth, who suffered from schizophrenia, committed suicide. In her suicide note she wrote that she was going into a parallel universe, to meet her father. His son Mark Everett became the famous rock star E, lead singer of the Eels. He described his father as distant, depressed, and mentally absent, and his own childhood as strange and lonely. Only his music saved him. But he also expressed sympathy for his father: "These guys, I don't think they should be held to subscribe to normal rules. I think that about rock stars, too."[30]

Hugh Everett's ideas about quantum physics were finally popularized in the 1970s by his advisor Wheeler and Bryce DeWitt, who had also worked with Wheeler. It was DeWitt who added the "many-worlds" label, a term that Wheeler never liked. The interpretation essentially states that every measurement results in a split of the universe. Every possible outcome of a measurement—or more generally of any physical process—is being realized, but in different parallel universes. If a guy chats up a girl in a dance club, there is always a universe where the two of them get happily married and remain in love until they die, but also another one where she tells him to back off, he has too much to drink, and he wakes up the next morning with a serious hangover. This very insight made me particularly nervous when I prepared to jump out of an airplane 4,000 meters above Oʻahu's north shore. After all, even if I survived in this universe, there are always countless universes where the parachute did not open. So somewhere one loses, every time. But somewhere there is also a parallel universe where Everett still lives happily together with his daughter.

The major advantage of the many-worlds interpretation, compared with the classical Copenhagen interpretation, is that no collapse of the wave function—which, in any case, is not really part of the theory—has to be assumed. Even after the measurement has been performed, both possible outcomes—like an electron at place A and an electron at place B—coexist, but they decouple, so that an observer who measures the electron at place A does not notice the alternative reality with the electron at place B. In contrast to the collapse of the wave function, this process of decoupling can be described within the formalism of quantum mechanics. Perhaps this process—so-called *decoherence*—is the only reason we witness so little quantum weirdness in our everyday lives. The drawback of the many-worlds interpretation, however, is that we have to give up the concept of a unique reality. The interaction of different parallel universes is suppressed after a measurement, but not totally lost. So even in our daily lives we don't reside in clearly defined conditions such as dead or alive. The parallel universes in which we and our fellow human beings experience totally different fates instead resonate as unobservable tiny admixtures of alternative realities into our universe. Thus the many-worlds interpretation exhibits the Parmenidic-neo-Platonic nature of quantum mechanics most clearly. According to this point of view, the unity of the different realities is not completely lost. It is actually possible to recognize the multiverse—the collection of all of Everett's parallel universes—directly as Parmenides's primeval One: the unity of the world the ancient Greeks felt they had lost in the charted modern world, and for whose reunification with the individualized ego they looked in the ecstasy of their mystery cults, in their Dionysian arts, or in the flush induced by psychedelic drugs.

The bizarre properties of quantum physics naturally inspired the fantasies of both journalists and authors. The parallel existence of different realities in quantum physics, for example, became the subject of a *Physics World* cover in 1998, which depicts a couple on the phone arguing as follows: "Oh Alice . . . you're the one for me"—"But Bob . . . in a quantum world . . . How can we be sure?" An even more radical take on the many-worlds interpretation can be found in Douglas Adams's *The Hitchhiker's Guide to the Galaxy*.[31] Whenever the extraterrestrial crackpot Zaphod Beeblebrox, double-headed and addicted to Pan Galactic Gargle Blasters, starts the Infinite Improbability Drive, his stolen spaceship gets located in all places in the universe simultaneously, and tiny probabilities are amplified. In the novel this allows the spaceship to travel faster than light, and also causes various strange incidents, such as when a threatening pair of rockets gets suddenly transformed into a dumbfounded whale and a flowerpot.

Finally, and now I am serious again, the many-worlds interpretation could protect time travelers from ludicrous paradoxes, and in this way make time travel a meaningful physics concept. But we'll get to that later.

4

Black Dots on a White Background
The Particle World

Physics is like surfing. Or like an LSD trip.

The first time I had a peculiar feeling of enlightenment was when I was staring onto a white surface sprinkled with tiny black dots. I was fourteen years old and, during an extensive study of my father's *Playboy* collection, I had developed a taste for modern and underground literature like Charles Bukowski, the Beat Generation, and Ernest Hemingway. While looking for new interesting books in my parents' bookcase, I stumbled upon Bertrand Russell's super-sized volume, *A History of Western Philosophy,* and, just a few pages into it, a captivating picture.[1] The picture was supposed to illustrate the philosophy of Anaxagoras, who lived in the fifth century BCE in ancient Ionia. Its caption reads, "Every single object contains a little bit of everything. What appears white on first sight actually contains traces of black when one takes a closer look."[2] Take for example opaque white, the purest white you can think of. If you put that under a microscope, you suddenly find black contamination. Such a picture can start you thinking: Does the whitest stone still contain black fragments? Fragments that are themselves made out of both black and white? Is matter infinitely divisible, or are there some smallest entities? Such questions have kept mankind pondering for more than 2,000 years. The ancient Greek philosopher Democritus actually proposed the existence of such smallest particles and christened them *atoms*.

What we call atoms in modern science are the building blocks of the isotopes in the periodic table. Pursue dissecting any of the items around you in your everyday life, and you end up with atoms. At some point you find a mixture of particles—molecules like H_2O when analyzing water, long chains of PVC molecules when dissecting your office chair, or elegantly winding double helices of DNA when probing within human or animal cells. Using the methods of chemistry you can dissect these molecules further into single atoms such as hydrogen and oxygen in the water molecule and carbon, hydrogen, and other atoms, such as chlorine, in synthetic materials or in organic molecules.

As scientists found out in the last century, however, the atoms themselves are by no means indivisible. The higher the energy applied, the smaller the constituents of matter that can be revealed. The atoms described in chemistry class are built of an atomic nucleus and the atomic shells in which electrons circle around the nucleus. But are there smaller parts? The atomic nucleus is itself a bound system made of protons and neutrons: The number of protons determines the element, while the number of neutrons determines the isotope. Finally, protons and neutrons are composed of quarks or, to be accurate, of the two lightest varieties of quarks, the *up* and *down* quarks. As of today the quarks, just like the electron, are considered to be *fundamental,* or *elementary,* which means not divisible any more. And with that we have entered the world of elementary particles. These tiniest and most fundamental building blocks of matter can be differentiated by various properties, which, according to the laws of quantum physics, can be altered only by leaps, or *quanta.* Scientists label these properties by *quantum numbers.* The most important quantum number of a particle is its *spin.* The known elementary particles behave just like little spinning gyroscopes, and the spin describes the strength of the rotation and the alignment of its axis. It is one of the paradoxes of quantum mechanics that point-like particles can possess spin. It helps to resolve this paradox if one realizes that a particle is, after all, nothing but an energy portion of an extended quantum wave. Consequently the spin is an absolutely real property. Atomic spins can set iron bars in motion because of the conservation of angular momentum, and reflection from the spins of atomic nuclei generates the MRI pictures that are used for medical diagnostics.

Wolfgang Pauli, a fellow student and close friend of Werner Heisenberg in Munich, who later in this book will make his appearance as the father of the neutrino, found out in 1924 that two identical particles with a *non-integer spin*— particles later called *fermions,* after the Italian physicist Enrico Fermi—can never occupy the same state of motion. In the end, it is this property that is responsible for matter having any dimension at all. Without the *Pauli principle,* the electrons in heavy atoms would collapse toward the highly charged atomic nucleus and cause those heavy atoms to be much smaller than lighter atoms. But the Pauli principle forbids two electrons with the same spin alignment from coexisting in the same energy state, which implies that they have to be moving in different orbits around the atomic nucleus. Consequently the heaviest atoms are not shrunken but maintain a size comparable to, or even larger than, light atoms. Moreover, this is the reason that two atoms repel each other and thus bestow objects with length, width, and height.

The building blocks of matter (electrons and quarks) are all fermions with spin-½, as are the neutrinos (Fig. 4.1). Since the first prediction of quarks by Murray Gell-Mann and George Zweig (both then at Caltech, although not collaborators) in the 1960s, physicists have found six species of quarks (*up, down, charm, strange, top,* and *bottom*—or, for short, *u, d, c, s, t,* and *b*), as well as the *leptons,* three electron-like particles with different masses (electron, muon, tau), each of which pairs up with an electrically neutral neutrino (Fig. 4.1). This

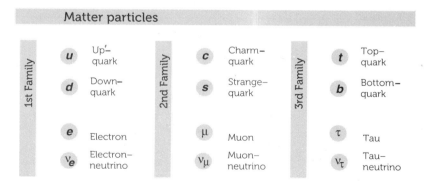

Figure 4.1. The elementary fermions, the building blocks of matter. There exist three generations, each with two quarks, one electron, and one neutrino.

threefold repetition of a known pattern (two quarks, with charge-⅔ and charge-⅓, one charged lepton with charge-1, and an electrically neutral neutrino) is referred to as the *three families* of the Standard Model. While the first family alone includes all fundamental constituents of the items of everyday life, the second and third families contain particles that were found in cosmic rays or at accelerator experiments. As of today, these twelve particles are considered the elementary essence of nature: Everything—really everything we can touch, smell, or taste—consists of these particles!

Moreover, every one of the elementary fermions exists with two possible spin alignments, left or right in relation to the direction in which the particle is heading. This property is called *helicity* (Fig. 4.2). A generalization of this observable— which remains unaltered for arbitrary observers and takes into account that a fast observer can overtake the particle and perceive an inverse direction of motion—is called *chirality,* or *handedness.* Finally, every particle possesses its own *antipar-*

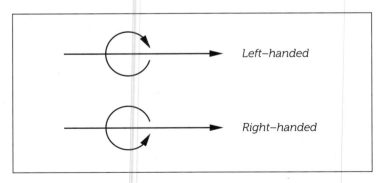

Figure 4.2. The two helicities that particles can possess. Chirality, or handedness, is a generalization of helicity that leads to the same result for arbitrary observers.

ticle, which denotes the corresponding particle with the opposite charge (for charged particles) and, in general, the opposite particle number; for example, the positron, which is a positively charged antielectron. When matter and antimatter meet, they can annihilate into radiation, but there exists no known large amount of antimatter in the universe. The reason for the preponderance of matter is one of the biggest puzzles of cosmology—a puzzle to whose solution neutrinos may contribute in a crucial way.

The different fermions in one family differ in various "charges" (not only electrical), which describe how particles react to external forces. For instance, electrically charged particles respond to electrical and magnetic forces. In the context of the theory of special relativity, electrical and magnetic forces can be understood as different aspects of electromagnetism. The electrical force describes the electromagnetism of charges at rest, while the magnetic force corresponds to the electromagnetism of moving charges. This is why electric currents are magnetic and why the magnetic properties of substances such as iron arise from circular currents and electron spins inside the materials.

The response to external forces is generally described as an interaction with a force field, such as the electromagnetic or the gravitational field. Electromagnetic waves describe the mutual oscillation of electrical and magnetic fields with which electromagnetic waves propagate through space. The different oscillation frequencies of electromagnetic waves are known to us as (with increasing frequency of the photon, the electromagnetic field quanta) radio waves, microwaves, infrared radiation, light, ultraviolet radiation, X-rays, and the gamma

rays emitted by radioactive materials. Again, quantum physics dictates that electromagnetic waves can propagate their
energy only in portions of finite quanta. Every force field thus
has a corresponding particle, a carrier transmitting the force.
A necessary condition is that these force carriers can be assembled together in common long collective wave trains.
Thus these particles cannot be fermions. The quanta of the
force fields are particles with an integer spin, so-called *bosons*
(after the Indian physicist Satyendra Nath Bose). The quantum of the electromagnetic field is the photon.

   In total there are four fundamental forces that act among
the elementary particles (Fig. 4.3): the strong force (acting only
on quarks), the electromagnetic force (acting on all charged
particles), the weak force (acting on all particles with left-
handed chirality), and finally the gravitational force (acting
on any kind of mass or energy). So far the force carriers of the
first three forces have been found:

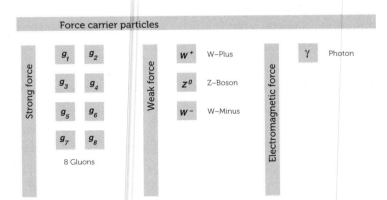

Figure 4.3. Elementary force carriers: the particles transmitting the
strong, weak, and electromagnetic forces.

- the *photon* of the electromagnetic field,
- the eight *gluons* of the strong interaction, and
- the three force-carrier bosons of the weak interactions, the W^+ W^-, and Z^0 bosons.

The gluons are responsible for holding together the quarks inside the protons and neutrons in the atomic nucleus and, as a residual effect, also for holding the nucleus together. The charge of strong interactions is called *color*, but bears no relation, of course, to the colors in our everyday lives. The name is based on the fact that, just as in the fine arts there exist three basic colors, there are also three color charges in nature (*red, green,* and *blue*), and that the three color charges, just as for the colors of actual light, add up to something neutral or white.

Among the elementary fermions, only the quarks carry color charge, but in contrast to the photons of electromagnetism, the gluons themselves are color charged (with one color and one anticolor each) and interact among themselves. This, combined with the number of gluons, is the reason for a curious phenomenon, namely that the color interactions become stronger with increasing distance. An important consequence is that in nature, only color-neutral objects can be found. These objects include so-called *baryons,* objects made up of three quarks, all carrying different colors, such as the proton *(uud)* and the neutron *(udd),* or *mesons* made up of a quark-antiquark pair, where the antiquark carries the anticolor of its quark partner. An important example for mesons is provided by *pions* ($u\bar{u}$, $u\bar{d}$, or $d\bar{u}$), where the bar denotes the corresponding antiquark). If one tries to dissociate one of these color-neutral compound states, the force between the quarks increases so quickly with the separation distance that the energy supplied is sufficient to create new quark-antiquark pairs out of the strong gluon field between them. These new particles combine with the fragments of the original compound particle in such a way that the resulting bound states are color-neutral again.

The weak interaction, mediated by the W^+, W^-, and Z^0 bosons, is the only interaction that can alter the particle type. It can, for example, transform an electron into a neutrino or a d quark into a u quark. This is called a change of the *flavor* of the particle. These force carriers have both electric and weak charge and can add or carry these charges in a flavor-changing interaction in order to transform one particle into another.

Unlike the massless photons and gluons, the bosons of the weak interactions possess relatively large masses, masses that in fact account for the weakness of the interaction. The basic weak interaction between force carrier and fermion is not *intrinsically* weak. What makes the weak interaction weak in practice is the propagation of the bosons themselves, since the low energies available in typical situations in our environment are insufficient to create a particle having such large mass. Accordingly, weak-interaction processes such as nuclear beta decay are only possible due to quantum fluctuations, a quantum effect based on Heisenberg's uncertainty principle. Let me discuss this principle for a moment. It forbids simultaneous accurate measurements of certain quantities: If you reconsider carefully last chapter's discussion of the double-slit experiment (Fig. 3.2), you will realize that by measuring the location of a wave by having it pass through a slit, one creates diverging circular waves behind the slit. Consequently, the component of its velocity in the direction of the determined position coordinate becomes uncertain. On the other hand, once you measure the velocity, the location becomes uncertain. A simultaneous exact determination of location and velocity is impossible. This phenomenon gave rise to various jokes, such as the claim that Heisenberg's tombstone carries the inscription "He lies here—somewhere" or the fictitious conversation of Heisenberg with a police officer: "No sir, I don't know how fast I was driving but I know exactly where I am." An uncertainty relation similar to the one between velocity and location exists between energy and lapse of time: In short time periods, the energy of a particle becomes uncertain. Weak interactions at low energies proceed then by making use of this energy uncertainty for short time lapses in order to create a W or Z boson as a short-lived fluctuation out of nothing. The short lifetime of this fluctuation makes such a process then so improbable that the resulting interactions between two fermions appear as weak.

As we shall see later, such quantum fluctuations play a crucial role in particle physics, ranging from the generation of neutrino masses to the birth of the universe itself.

The fourth fundamental force, gravitation, is not included in the Standard Model of particle physics. The reason for this is that, as of today, it has not been possible to develop a consistent theory of quantum gravity that would permit describing all forces of nature in terms of one unifying theory. We will meet the most promising candidate for such a theory— string theory—as well as some of its bizarre predictions, in Chapter 12.

It took more than ten years from their first attempts for Sheldon Glashow and Steven Weinberg of Harvard University—two researchers who had been classmates and competitors as students at Bronx High School of Science—together with Abdus Salam (whom we shall get to know in Chapter 6) and many others, to integrate these ideas into a consistent theory of particle physics, the Standard Model. This is perhaps the most successful theory in all of science: a theory that describes nature at distances as small as 10^{-16} centimeters and allows the prediction of properties of elementary particles such as the magnetic moment of the muon up to the eleventh decimal place, an accuracy that chemists, biologists, medical and social scientists, economists, and even condensed-matter physicists can only dream of.

The eventual icing on the cake was the discovery of the W and Z bosons in 1983 at the European laboratory for particle physics, CERN, in Geneva, Switzerland. The race for this discovery between the two experiments—Underground Area 1 and Underground Area 2 (UA1 and UA2)—was reminiscent of an epic clash of titans, battling with all possible means, fair and unfair, accompanied by assorted strokes of fate, and has been recorded in a thrilling book, *Nobel Dreams,* by Gary Taubes (a former Harvard physics student who used to write about boxing for *Playboy* magazine).[3] The hero of the drama, the UA1 group leader and later Nobel Prize winner Carlo Rubbia, eventually elbowed his way to success with outstanding skills, unlimited stubbornness, and ruthless vigor.

First he had to convince all skeptics—an almost Sisyphean task—that it actually would be possible to realize his and Simon

Van der Meer's ambitious project: to transform the existing Super Proton Synchrotron at CERN into a machine that accelerated not just protons in one direction but protons and antiprotons in opposite directions, eventually colliding them in a frontal impact. It was an idea originally conceived during a conference coffee break and was considered, at that time, "technically monstrous."[4] After the collider had finally been constructed, Rubbia's team had to fight, during data taking, with leakage of air into their detector, a water main break, and flooding. In order to reap the fruits of their labor after all the challenges, Rubbia supposedly made his result public hastily while, at the same time, urging his competitors to wait for more data and an unquestionable confirmation. Eventually he announced the discovery one day earlier than the group conducting the competing experiment, UA2. A few months later, in the fall of 1984, during a cab ride to the International Centre for Theoretical Physics in Trieste, he heard on the radio that he would receive the Nobel Prize and instantly wanted to share his pleasure with the driver. According to legend, the driver refused to believe he was driving the newly named Nobel Prize winner.

As impressive as the Standard Model appears today, hints have been accumulating in recent years that it is not the complete theory of particle physics—that apart from gravitation, there must exist new physics beyond the Standard Model physics, meaning new particles and interactions that need to be built into the theoretical framework.

The clearest of these clues have been found in neutrino physics. Grand hopes of uncovering the new physics now rest on

Table 4.1. The electron volt, the unit of energy and mass in particle physics		
Unit	Abbreviation	Numerical value
Electron volt	eV	$1\,eV$
Kilo electron volt	keV	$10^3\,eV$
Mega electron volt	MeV	$10^6\,eV$
Giga electron volt	GeV	$10^9\,eV$
Tera electron volt	TeV	$10^{12}\,eV$

new accelerators: the existing Large Hadron Collider (LHC) at CERN and the planned International Linear Collider (ILC), whose home, as of this writing, is yet to be determined.

Since the 1950s, physicists have used increasingly huge accelerators to produce the energy necessary to disrupt ever smaller particles and, in this way, to reveal their constituents or produce even totally new particles. Accelerators are nothing but vacuum tubes in which particles like protons and electrons and their respective antiparticles are accelerated up to nearly the speed of light and then caused to collide. The standard unit of energy in particle physics is the electron volt, or eV, which is the energy acquired by an electron accelerated through a voltage difference of 1 volt. Table 4.1 shows the abbreviations for the units of multiple electron volts as achieved at modern accelerators. The masses of particles can also be measured in electron volts, thanks to Einstein's mass-energy equivalence, and correspond to the minimum energy necessary to produce a particle of this mass.

The newest "magical" device, the LHC (Fig. 4.4), is set to beat all records. Its magnets are assembled in a tunnel twenty-

Figure 4.4. View into the Large Hadron Collider tunnel. With a circumference of 27 kilometers, the LHC, crossing four times under the French-Swiss border, is the largest particle accelerator of all time, a machine of unparalleled superlatives. (Courtesy Fermilab)

seven kilometers in circumference that crosses under the French-Swiss border four times, cooled down to minus 271 degrees Celsius—just two degrees above absolute zero and cooler than outer space. It is scheduled to accelerate protons up to 99.9999991 percent of the speed of light, in order to eventually collide them with a total energy of 14 TeV (trillion electron volts) inside one of its four detectors (Fig. 4.5), which are called ATLAS, CMS, ALICE, and LHCb. The required electricity corresponds to that necessary to power a medium-sized city such as Geneva.

Now that teething problems have been overcome (such as overheating caused by a baguette lost by a bird), new

Figure 4.5. The construction of the LHC detector called CMS. Visible is the onion-like structure containing different detector components that are sensitive to differently penetrating and differently charged particles and allow for an identification of the collision products. (Courtesy CERN)

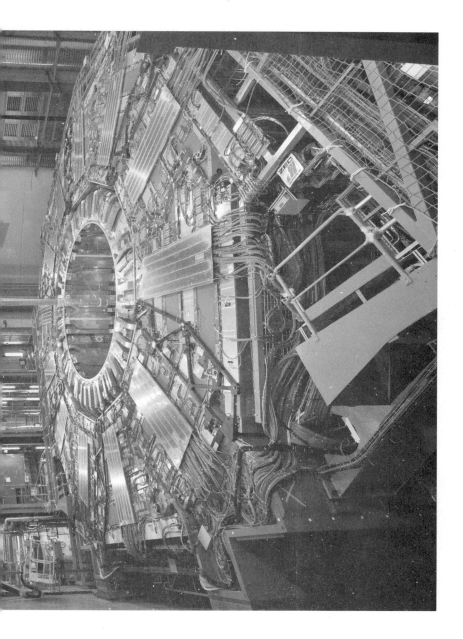

breakthroughs in particle physics are expected from the analysis of many thousands of collision products, such as the discovery of the Higgs boson, for which the first evidence was reported in July 2012; finding "supersymmetry"; and maybe even revealing extra dimensions—for example, by producing microscopic black holes. The prospect of the latter has already inspired several paranoiacs and crackpots to presage the end of the world. But even if we enter uncharted waters here, the LHC is harmless. What happens under carefully controlled lab conditions happens anyway every day above our heads when cosmic rays hit the upper atmosphere.

There is no law saying that quarks and leptons cannot be composed of even smaller particles, but so far there is no experimental evidence suggesting this. Moreover, with modern accelerators we have reached a realm where the energy necessary to disrupt the known particles is larger then the energies necessary to create new particle-antiparticle pairs from the vacuum. It is thus questionable whether the concept of further substructure makes sense at all. In the meantime, physicists have developed an even more radical idea of unification. They assume that at sufficiently high energies the different particles become identical. Such theories constitute the most ambitious hopes for the LHC, and they make use of a powerful concept of astonishingly aesthetic appeal: abstract symmetries. In this way physicists are following the path laid out by Plato more than 2,000 years ago.

5

Beyond the Desert
Symmetries and Unification

For forty years the tribes of Israel wandered through the desert, escaping slavery and genocide in ancient Egypt, before Moses led them to the Holy Land. When they arrived on the banks of the river Jordan, Moses climbed up Mount Nebo to die there. But before he died, he enjoyed a last glance into the country where his descendants were about to settle: a good land and large, abundant with grapes, figs, and pomegranates—a land flowing with milk and honey.[1]

Just like the tribes of Israel, particle physicists have been on a quest for forty years now, with physicists searching for hints of a theory of grand unification that traces all known particles back to a fundamental principle. They build increasingly gigantic machines, survey space with super-sensitive satellites, and push ahead farther and farther through the particle desert, the empty energy region where no new particles can be found—up to higher energies and back to earlier times in the history of the universe. The guiding theme that leads physicists and led the Israelites is the same: beauty. The holy land of particle physics is a realm of symmetries.

Symmetry is a word that is well known from art and aesthetics. Symmetries can be understood in a very simple way, mathematically, as properties of objects that remain unchanged under certain transformations. If, for example, you admire the beauty of a butterfly or of a Greek temple, that is at least partly due to the symmetry of the object. The image of a butterfly is mirror symmetric, since mirroring it along its central axis—meaning an exchange of left and right—yields back the original image (Fig. 5.1). The same holds for the temple, whose columns in addition exhibit symmetry under translation (Fig. 5.2). And very similarly, a snowflake possesses a six-fold rotational symmetry, since a rotation by 60 degrees results in the original image (Fig. 5.3).

In the realm of physics, the notion of symmetry is closely aligned with the name Noether.[2] Emmy Noether was a Jewish

Figure 5.1. Mirror symmetry of a swallowtail, one of the most impressive butterflies, found in both North America and northern Europe. (Butterfly image courtesy Ullstein Bild/The Granger Collection)

Figure 5.2. Mirror and translation symmetries of the Parthenon in Athens. (Parthenon photo courtesy Ullstein Bild/Prisma/VC Ross)

Figure 5.3. Rotational symmetry of a snowflake. (Snowflake image courtesy Yoshinori Furukawa)

German mathematician of the late nineteenth and early twentieth centuries. Her short life was anything but easy. She was subjected to discrimination, exclusion, and even persecution: Being a woman, she was at first not allowed to apply for a college education. Later she became a well-known contributor to the field of differential invariants and had the support of the important mathematicians David Hilbert and Felix Klein. Nevertheless the *Habilitation,* the traditional German prerequisite to apply for professor positions, was initially denied her. No exception was made for her, and female candidates in general were accepted only after World War I.

Three years after she finally accomplished her *Habilitation,* she lost her entire small fortune—which had supported her humble life—in the inflation of the Weimar Republic. After returning from a visit to Soviet Russia, she finally obtained a visiting professorship in Frankfurt, but only three years later she had to flee the country, immigrating to the United States to escape from the Nazi terror against Jews. There she died less than two years later of breast cancer. Her brother, who fled to the Soviet Union, was murdered there shortly afterward during Stalin's "great purge." It is thus even more impressive that Emmy Noether, of all people, had the gift to see the essence behind a set of mathematical formulas, to see the symmetry behind the equations. Remarkable that it was Emmy Noether who opened the eyes of physicists to what for many of them would become the principle of the beauty of nature.

Ever since Newton, physics has been characterized by conserved quantities. Even if you didn't know anything at all about a mechanical system, you still could be sure that quantities such as energy and momentum (the product of mass and velocity) wouldn't get lost. It was Noether's spectacular discovery that these conserved quantities were inseparably linked to the symmetries of physics. Consider for example a ball rolling on a plane. If the plane is flat, the ball will keep its velocity and thus its momentum. There exists a conserved quantity. But if the plane is askew or contains bumps or craters, this will accelerate or decelerate the ball, and the momentum is not conserved (although the total momentum of ball and plane remains conserved). The plane possesses a translational symmetry in the first case, while in the second case no such symmetry exists (Fig. 5.4). What Noether realized was that the reason for the conservation of momentum was nothing but the translational symmetry of the environment

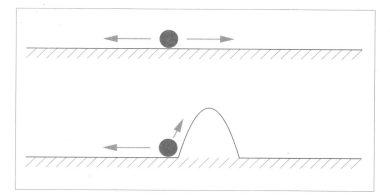

Figure 5.4. A flat plane after a horizontal translation looks the same as before, and the momentum of a ball rolling along such a plane is conserved (upper picture). An inclined plane or a plane with bumps or craters changes under horizontal translations. A rolling ball will then be accelerated or decelerated, which means its velocity is changing (lower picture).

or, more exactly, the translational symmetry of the physical laws governing the system under consideration. And she saw even more: that every symmetry—even totally abstract ones, as we shall see later—requires a conserved quantity.

Just as translational symmetry underlies conservation of momentum, the conservation of angular momentum is related to symmetry under rotations, and the conservation of energy rests on the symmetry of physics at different times, meaning under time translations.

This link between conserved quantities and symmetries is of crucial importance in modern physics. Known groups of elementary particles, such as the quarks, can be classified according to their quantum numbers. Surprisingly, the quarks can then be arranged at the corner points of polygons that have maximally symmetrical shapes. If a quark is missing from one of the corner points, the corresponding particle with its quantum numbers can be predicted. But this by no means exhausts the role that symmetries play in modern physics.

Following Noether, the importance of symmetry consideration in physics can barely be overstated—from Einstein to modern particle physics. And this brings us back to the eighteen-year-old Werner Heisenberg and his reading break in the rain pipe. At least from that instant on, particle physics followed a Platonic program. In fact, the biggest dream of particle physics can be understood as the search for a mathematical formulation of the part of Plato's *Timaeus* that so captivated Heisenberg: "Since there exist transitions between the elements water, fire, earth, and air, it makes no sense to assume that they are real; there rather is a primordial substance, which becomes one of the elements in one of its appearances only."[3]

This unity that Plato and Heisenberg dreamt of, as do most contemporary particle physicists, can be reduced to a simple concept: an all-embracing symmetry. When he was sixty-eight years old, just seven years before he died, Heisenberg gave a speech on the occasion of a celebration honoring Alexander von Humboldt. "The unity of Nature can be understood today in the following way: All—even the most dissimilar—phenomena can be reduced to the same basic structures. And just a few fundamental symmetries and basic structures are sufficient to build up by repetition and by collective shaping the infinitely complex play of the observable world."[4]

In 1932 Heisenberg translated the idea of the Platonic primordial substance almost literally into the framework of particle physics. What for Plato were the elements water, earth, fire, and air correspond in modern physics to the elementary particles. Back then, at the dawn of particle and nuclear physics, Heisenberg introduced a symmetry relation between different particles that became an overarching guiding principle for the way particle physicists think today. Heisenberg realized that proton and neutron have almost the same mass, and he knew that they can be transformed into each other by nuclear beta decay and that, apart from their different electrical charge, they behave in very much the same way. Just as the butterfly and the temple are symmetric under an exchange of left and right, particle physics seemed to be symmetric under an exchange of proton and neutron. Consequently Heisenberg postulated that proton and neutron are actually just the same particle, a so-called *isospin doublet*. And this isospin doublet can appear as a proton or neutron—like the two sides of one coin or Plato's appearances of the "primordial substance." In this connection, the isospin is Noether's conserved charge and the name *isospin doublet* originates from the analogy of an electron in the atomic shell that can align its spin parallel or antiparallel to the angular momentum of its trajectory. Whether it has its spin pointing "up" (right spin) or "down" (left spin), we have one particle assuming one of two possible states. In summary, Heisenberg understood the proton and neutron as two sides of one coin.

But with this the relevance of the symmetry concept is by no means exhausted. Soon physicists asked themselves whether theories could make any sense where the exchange of protons and neutrons depends on location. For instance, one could exchange all protons with neutrons in New York, let all protons remain protons in Heidelberg, and transform the protons in Tokyo into some superposition which is half proton and half neutron. Would such a theory still correspond to the original one? The surprising answer is yes—but only if the theory contains forces. Such theories are called *gauge theories,* and they provide the mathematical underpinning of all modern particle theories. The symmetry here not only determines which kinds of particles can exist, it also dictates what forces act among them.

Naturally one can carry this approach to extremes, and needless to say, physicists have done so. In the 1970s, Howard Georgi and Nobel Prize winner Sheldon Glashow, and later Harald Fritzsch and Peter Minkowski at the California Institute of Technology, generalized these ideas as follows: If one understands proton and neutron (or, as has been found out later, neutrino and electron) as two states of one doublet, as "two sides of one coin," would it not be possible to understand all known elementary particles as different states of one primordial particle? While the Glashow-Georgi theory (called *SU(5)-GUT*) still needs two primordial particles, the Fritzsch-Minkowski theory (called *SO(10)-GUT*) realizes the concept using just a single entity.

The Platonic program following in Heisenberg's footsteps was accomplished in the sense that all quarks and leptons could be understood as the "16 sides of one coin"—as the sixteen states or *16-plet* of a single primordial particle. So-called *leptoquark* force carriers can then transform the sixteen states into each other.

But that's still not everything: To follow this spirit to the end one should not only *unify* all material particles—the different forces should also be depicted as different aspects of a single force. This idea carries the name *Grand Unified Theory* or simply *GUT*. The concept epitomizes such a breathtaking elegance that it bears an irresistible attraction for many physicists, irrespective of its difficulties. The ultimate answer to all possible questions would no longer be "42," as in Douglas Adams's science fiction parodies;[5] it would be "This is a result of the gauge symmetry." For many years now, CERN's chief theorist, John Ellis, has liked to wear a T-shirt that says "no GUTs no Glory" when he shuffles through the corridors of the world's biggest particle research center. And the difficulties are anything but small.

Unfortunately for the concept of unification, the masses of elementary particles measured in experiments are far from equal. Also, the effects of the electromagnetic, the weak, and the strong interactions differ hugely. Moreover, the strength of the various interactions depends in different ways on the separation distance of the particles attracting or repelling each other: While the strength of the strong interaction grows with increasing distance, the strength of the electromagnetic interactions decreases. The symmetry of all forces and particles as imagined by Plato, Heisenberg, Glashow, Georgi, Fritzsch, Minkowski, and many others is only imperfectly realized in nature—it is broken!

The concept of an all-embracing symmetry can be saved, however, since particle masses and forces depend on energy. The reason for this is that an electrically charged particle will "polarize" the vacuum by generating particle-antiparticle pairs: Positive and negative charges pop out of nothing via quantum fluctuations. The negative fluctuations then get attracted by the positive charges and the positive fluctuations accumulate around the negative charges, thereby screening their effect (see Fig. 5.5). At higher energies or at smaller distances, this charge cloud can be penetrated more and more deeply, and the force then increases in strength. What is fascinating is that this effect is actually big enough to be observable: It is responsible for the various forces becoming more alike with increasing energy! Thus it is really possible that a GUT symmetry is realized at very high energies, and that at low energies we only experience an imperfect image of the underlying symmetry (Fig. 5.6).

But will it be possible to prove that such a hidden symmetry actually exists? (Gravitation, being described as the curvature of space-time, is qualitatively so different from the other

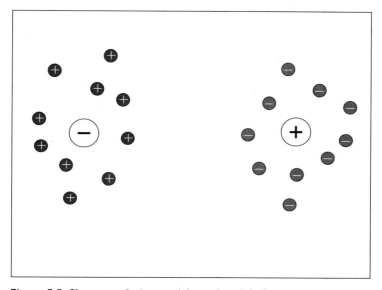

Figure 5.5. Charges polarize particle-antiparticle fluctuations in the vacuum and thus get screened. Charged particles then act not according to their original but according to their screened charges. As a consequence, force fields become weakened or amplified.

forces that its unification with these other interactions requires special efforts and care: see Chapter 13.) To reach the energies where a Grand Unification would actually be realized, particle physicists would need a circular accelerator with a circumference of a trillion kilometers, which is almost a million times as big as the entire solar system. What can be found in between the range of validity of the Standard Model and that of a Grand Unification, a realm extending over thirteen orders of magnitude, or a factor of 10 trillion in energy, in which originally no new physics was expected, is denoted as the *particle desert.*

The most promising prospect for receiving signals from beyond the desert seemed to be the search for the decay of the

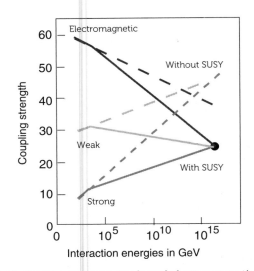

Figure 5.6. Unification of strong, weak, and electromagnetic force at a high energy scale into a Grand Unified Theory.

proton. As discussed above, the leptoquark particles existing in GUT theories allow for the transformation of different particles into each other, for example quarks into leptons. Consequently a proton, which is made out of quarks and is totally stable in the Standard Model, can decay into lighter mesons and leptons. Eventually nothing would be stable anymore, even we ourselves would decay in the course of time into lighter particles, albeit with a gigantic half-life, about ten million billion billion billion times longer than our typical life span. Nevertheless such decays can in principle be observed. Since the 1980s, bigger and bigger experiments have been constructed to look for proton decay. In many of them, electronic eyes, so-called *photomultipliers,* scanned the interior of a huge water tank for hints of a decaying proton. While these experiments have so far obtained no evidence for decaying protons, on the bright side they have become the most important instruments for researching neutrino masses and, in this way, have delivered at least an indirect hint about the physics of Unification.

So physicists dream about the physics beyond the desert—a dream for which my PhD advisor Hans Volker Klapdor-Kleingrothaus and I tried to create some momentum by initiating the conference series Beyond the Desert in 1997. In our research group it became common to decorate the slides for our presentations with camels struggling their way toward an oasis in the desert. My camels always wore sunglasses, carried big beer mugs, and showed off big grins to convey that neutrino physics is perfectly prepared to find its way through the desert. For neutrinos are special particles and could really turn out to be keys to the universe, as we shall see in Chapters 10 and 11.

But why should the various interactions evolve so differently out of a totally symmetric constellation? And why should the material particles that we observe be so unequal, with different interactions and different masses? How does the diversity in the universe arise if everything is built up only from the various states of one single primordial particle and the different aspects of one primordial force? The answer to these questions is a concept called *symmetry breaking,* and ironically this concept leads, if you think it out, to an even higher degree of symmetry than is incorporated in the GUT theories, namely to something called *supersymmetry* (or SUSY). And as we shall see, the concept of supersymmetry not only makes it possible to solve some of the conceptual problems of the Standard Model and of GUT theories, it also predicts the existence of new particles in the middle of the particle desert. As the SUSY theorist Howie Baer declared in his summary talk at the first Beyond the Desert conference, it makes the desert bloom.[6]

From Symmetry Breaking to Supersymmetry

What is the origin of the world's diversity, the endless complexity of our surroundings, if everything is made out of a single primary type of particle and held together by a single primary type of force? The concept that solved this problem was first discovered in solid-state physics, and it is reminiscent of an old Irish legend (as found in the description of the piano piece *Voice of Lir* by Henry Cowell): "Lir of the half tongue was the father of the gods, and of the universe. When he gave the orders for creation, the gods who executed his commands understood only half of what he said, owing to his having only half a tongue; with the result that for everything that has been created there is an unexpressed and concealed counterpart, which is the other half of Lir's plan of creation."[1]

In this way a blemish came into the world—and very similarly is symmetry breaking understood in modern physics.

In 1964 Peter Higgs in Edinburgh, Scotland, submitted a two-page manuscript to the journal *Physical Review Letters,* which was to make him a celebrity. In this paper he proposed a new field, now called the Higgs field, and as a quantum of this field a new boson (the Higgs boson), which later even earned the title "God particle." Actually Higgs is not famous for his diligence. It was only the second paper submitted by this thirty-five-year-old lecturer at the University of Edinburgh, and in his entire scientific career he published no more than five papers. These days, with less than twenty papers one doesn't even get considered for a faculty job. But if the Higgs boson has really been found at the LHC, Peter Higgs could easily become a Nobel Prize winner. On July 4, 2012, this gratification finally seemed to be within Higgs's and his fellow discoverers' grasp: In the overcrowded CERN auditorium, in a scene reminiscent of Rubbia's triumph twenty-nine years earlier, CERN General Director Rolf-Dieter Heuer exclaimed, "This is an historic day, but we are only at the beginning."[2] His audience, some of them moved to tears, included Higgs as well as François Englert, Gerry Guralnik, and Carl R. Hagen, who also contributed to the Higgs mechanism. And later, in a press conference, Heuer continued, "As a layman I would say, I think we have it! We have a discovery—we should state it. We have a discovery. We have observed a new particle consistent with a Higgs boson." Whether the particle found is really the Higgs, either in its Standard Model version or in some extended theory, remains to be proved, but chances are high that the breaking of symmetry and the origin of mass are now finally being understood.

The topic of Higgs's famous paper is a mechanism called *spontaneous symmetry breaking,* which allows physicists, in view of the dissimilar forces and particles, to keep on speculating about unifying them in GUT theories. But apart from that it plays an even more important role: Only the Higgs mechanism allows for particles to have mass in the Standard Model in the first place.

Spontaneous symmetry breaking occurs if the fundamental equations of a system obey a certain symmetry, but the so-called *ground state,* the state with the lowest possible energy, does not. The most prominent example for this phenomenon is a ball in a mountainous landscape (dubbed a *potential*) that resembles a Mexican hat, the so-called *Mexican hat potential* (Fig. 6.1). The hat displays obvious mirror and rotation symmetries. A ball rolling around in the landscape finds its final position not at the center but in the brim. Therefore its rest position, or ground state, is symmetric neither under mirroring nor under rotations. The symmetry is said to be broken for the ground state. An important aspect here is the energy dependence of the problem: As long as the energy of the ball is large enough, it can override the crest in the middle

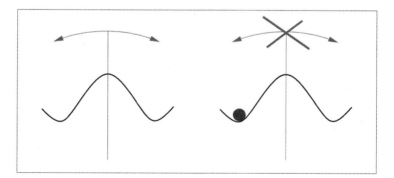

Figure 6.1. Spontaneous symmetry breaking in the Mexican hat potential. The landscape shown at the left has a mirror symmetry, but as soon as the ball assumes its rest position (right) the symmetry is broken.

and assume any position inside the hat. At high energies the symmetry of the basic equation is therefore restored.

This mechanism plays a decisive role in many areas of physics. For example, one can understand a magnetizable piece of iron as composed of many tiny molecular magnets corresponding to individual atomic spins. As long as all molecular magnets are aligned, the iron is magnetic. If, however, they point chaotically in random directions, the individual magnetic moments cancel (Fig.6.2 left). At high energies the magnets continually interact, or "scatter" from one another, and there is no net alignment. Consequently the magnetic field of the piece of iron is zero; it does not point in any direction. This means that there is no preferred direction, and there exists a rotational symmetry. But if one lowers the energy of the molecular magnets, for example by cooling the piece of iron, the interactions are weakened (less movement and thus less "scattering") and the mutual magnetic attraction begins slowly to align them. At a low enough temperature, the molecular magnets will point in the same direction and their magnetic

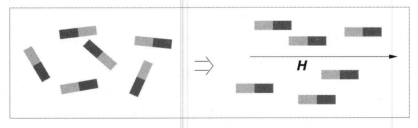

Figure 6.2. Alignment of molecular magnets in a piece of iron. At high temperature (left) the magnets are disordered, the total magnetic field is zero, and a rotational symmetry exists. At low temperature (low energy, right) the magnets get aligned and add up to a resulting non-zero field. The direction of the magnetic field breaks the rotational symmetry.

moments will align, bestowing on the piece of iron an overall magnetic field in some particular direction: The rotational symmetry is broken (Fig. 6.2 right). This happens despite the facts that the fundamental laws of physics specify no specific direction and cannot predict in which direction the emergent magnetic field will eventually point.

Higgs—and almost simultaneously Robert Brout and François Englert, as well as Gerry Guralnik, Carl R. Hagen, and Tom Kibble—applied this symmetry-breaking mechanism to particle physics. As we know today, the universe in its early stages was a hot plasma of highly energetic particles. Higgs assumed that during these early stages—that is, at high energies and high temperatures—all particles obeyed a sweeping symmetry: They could be considered to be differing states of a single primordial particle. This situation changed when the universe expanded and cooled down. Just as the moisture in warm air condenses on the surface of a beer bottle that has just been fetched from the fridge, the Higgs field condenses in the vacuum of the cooling universe. Just as the ground state of the iron piece now lacks rotational symmetry, the ground state of the Higgs field no longer respects the symmetry among the different particle species. As soon as the Higgs field assumes its rest position, meaning it condenses in the vacuum, neutrinos and electrons do not look at all alike anymore. They appear totally different. The mechanism of spontaneous symmetry breaking can explain why we do not observe the perfect symmetry of a GUT theory in nature, but only an imperfect reflection. It's rather like the people in Plato's allegory of the cave, who perceive only the shadows of original objects being projected onto the wall of the cave, and who consider these shadows as reality. The true forms of reality, which are responsible for the shadows—Plato's "Ideas" or "Forms" on which all perceptions are based—remain elusive. In Plato's *Republic,* the path to cognition lies in the realms of philosophy.[3] For the hidden symmetries of the particle world, it lies rather in quantum field theory.

It may sound paradoxical that the vacuum should be filled with something—the Higgs field—when "vacuum," almost by definition, means emptiness. But an empty vacuum is an alien concept in quantum field theory, where quantum fluctuations permanently pop in and out of existence. Or as Higgs himself expressed it: The vacuum is not exactly nothing.

What is considered to be the vacuum in quantum field theory is just the state in which the fields described in the theory possess the lowest possible energy: the ground state. And what is characteristic for the Higgs field is that it has its lowest possible energy in a state where the field strength does not vanish.

But what is the link between spontaneous symmetry breaking and the generation of masses? Within the Higgs mechanism the particles acquire their masses by rubbing themselves on the Higgs field filling the vacuum. But why is this necessary at all? And why is symmetry breaking a prerequisite for mass generation within the Standard Model?

This leads us to a crucial problem of the Standard Model—namely the question of how particle masses can be accommodated in the theory—and to a restless night flight over the Atlantic Ocean, during which a young man from a tiny village in what is now Pakistan broods over this very question.

Abdus Salam was fourteen years old when he scored the best result ever recorded in the qualifying exam of the University of Punjab. Afterward, he rode back on his bicycle to his village, where the people were standing on the road and cheering. Six years later he left Punjab with a fellowship for the University of Cambridge. Salam considered himself in the tradition of the medieval scientists, who left their flocks of sheep behind in northern Europe to join the great Arab universities of Toledo and Cordoba in order to bring back modern science to their home countries. By the time Salam was thirty years old, he was not totally unknown in the small circle of quantum field theorists. Nevertheless he had little money and so was happy when someone arranged a seat for him on a US Air Force transport flight for his trip back to Europe from a conference in Seattle. The overnight flight was very uncomfortable, and while he sat sleepless among soldiers and crying children, he mulled over the lectures he had just attended, where C. N. Yang and T. D. Lee described their suspicion that the law of left-right symmetry could be violated in weak interactions: Yang and Lee sensed that left-handed particles and right-handed particles would interact differently in weak interactions. And another memory came back to Salam: a question that Rudolf Peierls—a distinguished scientist, student of Heisenberg and Pauli, and, as the head of the British nuclear weapons program, Commander of the Most Excellent Order of the British Empire—had asked Salam during his PhD exam. The question, to which Peierls admitted willingly he didn't know the answer himself, was: Why is the neutrino massless? Could that question be related to the thinking of Yang and Lee?

Today we know that the question was wrong: Neutrinos are not massless after all. But the answer Salam came up with was significant not only for neutrinos; it would change the entire way particle physicists thought about mass generation in general.

What Salam knew was that the neutrino had to be massless if it had no right-handed partner. Consequently, since there exists no right-handed neutrino in the Standard Model, this would mean the neutrino had to be massless. But why is this the case? What if there were in fact a right-handed neutrino?

We remember that a particle with a counterclockwise, or left-handed, helicity could be overtaken by an observer and would then appear to be right-handed, that is, rotating clockwise. Special relativity imposes a speed limit here, though (see Chapter 13): The energy that is necessary to accelerate an object up to the speed of light grows without limit if the particle has mass. Only massless particles such as the photon can travel at the speed of light; a massive particle will never reach it, no matter how much energy is spent to accelerate it. An observer overtaking the particle is thus only possible if the particle does not travel at the speed of light—that is, if the particle possesses mass. Massless particles, on the other hand, always move at the speed of light, and their sense of rotation, their helicity, remains the same for all observers. Translating this context now into the observer-independent language of chirality, one finds that the mass of a particle can trigger a chirality-changing transition, or even more: that the existence of such a transition and the property of possessing a mass are equivalent. This explains why the neutrino has to be massless in the Standard Model: If there exists no right-handed neutrino, then no chirality-changing transition from left to right is possible, and consequently the neutrino has no mass. The violation of mirror symmetry in the Standard Model expresses itself in the observation that weak interactions couple to left-handed particles only, but it doesn't have to imply that a right-handed neutrino with no interaction at all (other than gravity or possible Higgs exchange) does not exist.

During his sleepless night far above the Atlantic Ocean, Salam realized even more: Even if the right-handed neutrino were to exist, the mass of the neutrino would still be forbidden, as a direct consequence of the violation of mirror symmetry. If the left-handed neutrino interacts with the weak force, Salam reasoned, it also has to carry the charge of the weak interactions, the isospin. But the right-handed neutrino does not interact, and thus does not carry any isospin. Now, as charges are always conserved in any circumstances, that would mean that transitions from left-handed to right-handed—exactly those transitions that correspond to neutrino masses—are impossible.

When he arrived in England the next morning, Salam was upbeat. He hurried off the plane, raced to his office in Cambridge to perform a few fast calculations, and then boarded a train to meet Peierls in Birmingham to ask him whether he would accept this answer to his PhD question. But Peierls disappointed him: He wouldn't believe in any violation of left-right symmetry and he "would not touch such ideas with a pair of tongs."[4] Salam pondered desperately whom else he could ask, and he came up with Pauli in Zurich. So Salam travelled to CERN in Geneva and handed his paper to one of Pauli's collaborators, only to get it back a day later with the message, "Give my regards to my friend Salam and tell him to think of something better."[5] Now Salam was really frustrated. But luckily, very soon afterward, Chien Shiung Wu, whose office was

only a few floors above the offices of Lee and Yang at Columbia University, measured the angular distribution of the radiation emitted by the beta decay of cobalt 60. She demonstrated that mirror symmetry is indeed violated in weak interactions. After that, Pauli sent Salam his apologies. Now
Salam felt encouraged to also send Pauli his later work, in
which he generalized the concept of left-handed couplings to
the charged leptons. But Pauli's reaction was no better than the
first time: He wrote to Salam that his aim for universality was
"pure nonsense."[6] Pauli actually had a very concrete criticism.
For massive particles such as the electron or its heavier siblings, muon and tau, the distinction between left and right
constituted a serious problem: They all had to be massless now!

As I argued before for the neutrino, the isospin of the left-
handed particles would be lost in the transition between left
and right—exactly that transition corresponding to the particle mass—whenever they were being transformed into the
right-handed particles, which do not interact weakly and thus
don't have any isospin. All particles would be massless unless
the isospin charge could somehow escape.

And here again the Higgs field comes into play: Within the
vacuum, the Higgs field itself carries isospin charge, and thus
can acquire or shed isospin charge by absorbing or emitting
quanta whenever fermions rub against the Higgs (see Fig. 6.3).

With this piece added, the mechanism of spontaneous symmetry breaking is complete: It is possible to endow all particles in the Standard Model with masses and, at the same
time, to unify electromagnetism and the weak interactions
into the electroweak force. For this theory Salam, together
with Glashow and Weinberg, received the 1979 Nobel Prize.

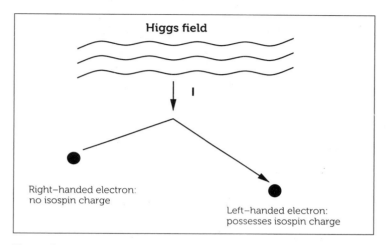

Figure 6.3. A particle rubbing against the Higgs condensate within the vacuum. In this way the particle acquires its mass and can lose or gain excessive or missing isospin charge from the vacuum.

Moreover, it is clear now how this concept can be generalized into the Grand Unification of the GUT theories. At that point, however, the question of neutrino masses comes back into play, as we shall see later.

Apart from this, the paradigm of symmetry seemed to be exhausted—until physicists started to calculate the mass of the Higgs boson. It cannot be much larger than the masses of the $W^{+/-}$ and Z^0 bosons for the mechanism of spontaneous symmetry breaking in the Standard Model to work properly. However, quantum fluctuations in which the Higgs particle metamorphoses temporarily into other particles contribute to the Higgs mass. And these quantum fluctuations make the Higgs boson very heavy, as heavy as the GUT energy, meaning too heavy by a factor of 10^{14} unless all fluctuations accidentally cancel. This problem is know as the *hierarchy problem* ·

(why is the GUT energy so much larger than the masses of the W and Z bosons?) or *fine-tuning problem* (why does the cancellation work so accurately?). So far this problem is not solved, but physicists have a favorite candidate for the solution, and ironically it means even more symmetry: the so-called *supersymmetry*, or, for short, *SUSY*.

If GUT theories carry the Platonic program to extremes, then SUSY shoots it right through the roof. Supersymmetry means that not only particles and forces are unified among themselves, but that matter and forces are unified with each other. Or more concretely: Matter and force particles are understood now as different states of a single *superfield*. This scenario, which was originally developed in the context of string theory (Chapter 12), is of impressive mathematical elegance: Matter and force particles form pairs, meaning that every elementary fermion with spin-$\frac{1}{2}$ gets a corresponding boson as a SUSY partner, and every force carrier as well as the Higgs boson itself gets a fermion as a SUSY partner (see Table 6.1).

The solution that SUSY offers for the hierarchy problem is simple: While quantum fluctuations of both bosons and fermions contribute to the Higgs mass, they enter the equations with a different sign. If now for every fermion a corresponding boson exists, which in all other respects is identical to its fermion partner, the contributions from the quantum fluctuations of both particles cancel exactly. Consequently the Higgs boson can have the desired mass in order to realize the spontaneous symmetry breaking in the Standard Model.

Back in the early 1970s, Pierre Ramond of the University of Florida, and, in the Soviet Union, Dmitri Volkov and Vladimir Akulov, as well as Yuri Gol'fand and Evgeni Likhtman,

Table 6.1. Standard Model particles and their SUSY counterparts	
Particle	SUSY Partner
Quark (s = $^1/_2$)	Squark (s = 0)
Electron (s = $^1/_2$)	Selectron (s = 0)
Muon (s = $^1/_2$)	Smuon (s = 0)
Tau (s = $^1/_2$)	Stau (s = 0)
Neutrino (s = $^1/_2$)	Sneutrino (s = 0)
Photon (s = 1)	Photino (Neutralino) (s = $^1/_2$)
Gluon (s = 1)	Gluino (s = $^1/_2$)
W boson (s = 1)	Wino (Chargino) (s = $^1/_2$)
Z boson (s = 1)	Zino (Neutralino) (s = $^1/_2$)
Higgs boson (s = 1)	Higgsino (Neutralino) (s = $^1/_2$)

had independently tried out some supersymmetric models, albeit in a rather formal context:[7] string theories with one space dimension and abstract generalizations of the algebra of the mathematical structures of relativity. In 1974, Julius Wess and Bruno Zumino at Karlsruhe University developed a four-dimensional supersymmetric theory that later was extended into a realistic particle theory by Zumino and Sergio Ferrara as well as by Howard Georgi and Savas Dimopoulos. A fascinating aspect revealed by these works is that SUSY constitutes the only possibility for describing the inner symmetries of particles (such as invariance under the change of charges) in a consistent mathematical framework together with a symmetry acting in space that exceeds the symmetries of space itself and acts on the rotational properties of the particles.

Symmetries bear—as we have seen already several times—a magical fascination for particle physicists. The only problem

with the concept is that so far not a single force-matter particle pair that fits together has been found. Thus, SUSY forces
physicists to complement the particle spectrum of the Standard Model with the SUSY particles—and thus double the
number: a dreadful thing for the fetishists of rationalization
but a feast for experimental physicists who search for new
particles. Moreover, the SUSY partners have to be more massive than the Standard Model particles; otherwise they would
have been found long ago. Thus SUSY, which technically requires identical properties and thus equal masses of particles
and SUSY partners, has to be broken. The energy where this
breaking occurs then determines the order of magnitude of
the masses of the SUSY particles. In order for the stabilization
of the Higgs mass to work, the masses of the SUSY particles
cannot be larger than about 1 TeV. And this is exactly the energy range probed right now at the LHC.

Most particle physicists are fascinated by SUSY, the confirmation of which, with the discovery of the first SUSY particle,
is—next to the discovery of the Higgs boson—the biggest
hope in the search for new physics at the LHC. This is partly
because of the sheer beauty of the concept. Einstein once said
that if general relativity were not realized in nature, he would
have felt sorry for the dear Lord. A similar attitude toward
SUSY is common, but it is not the only reason for the enthusiasm of the particle-physics community. After all, SUSY can
solve very concrete problems:

- *The unification of forces in GUT theories.* If one
  calculates how the strengths of the strong, weak,
  and electromagnetic forces change with energy

(see Chapter 5) one finds that they do not converge to a single point at high energies. This is a problem for GUT theories, for which exactly this convergence at the GUT scale is predicted. The energy dependence of the strengths of the forces depends on vacuum fluctuations that screen the charges of the various forces at larger distances, corresponding to lower energies, and thus depends on the number of particles that can fluctuate within the vacuum. If energies in the TeV region make it possible to produce SUSY particles fluctuating in the vacuum, these new particles affect the evolution of the strengths of the forces in such a way that they converge at a single point at high energies and thus make possible a consistent GUT theory.

- *The dark-matter candidates.* If one assumes that SUSY particles can be created or annihilated in pairs only, meaning that so-called *R-parity* is conserved, then the lightest SUSY particle cannot decay and is stable. One can show that the abundance of such a particle—typically the lightest neutralino—is of the right order of magnitude to make up for the dark matter in the universe (see Chapter 10). But also, in theories where SUSY particles can all decay into Standard Model particles, the SUSY partners of the hypothetical force carrier of gravitation, the gravitinos, would be candidates for dark matter.
- *Quantum gravity.* SUSY seems to be a necessary ingredient of consistent quantum gravity theories.

The reason is that SUSY, once promoted to a
gauge theory, automatically contains relativity as
well. In particular, string theories (see Chapter 12)
predict supersymmetry, but they do not determine the masses of the SUSY particles.

With the mechanism of spontaneous symmetry breaking
and the Higgs boson, the Standard Model gets completed in
an elegant way. For the so-far unanswered questions, such as
What is the dark matter in the universe? and How can gravity
be described in terms of a quantum theory? GUT theories
and SUSY offer compelling and aesthetically appealing candidates for a theory beyond the Standard Model. But how can
these theoretical constructs be confirmed by experiment? Is
there really any physics beyond the Standard Model?

Yes, there is. And it was found in the study of a particle
that is very different from all the other leptons and the quarks:
the neutrino.

Birth of an Outlaw
The Neutrino

Among all elementary particles, the neutrino is surely the most exotic: Although sixty billion neutrinos pass every second through each square centimeter of the surface of the earth—as well as through every human body, or any other object—these particles penetrate through us and the whole earth's interior as if through thin air. And although every single neutrino weighs less than one millionth the weight of the tiny electron (which itself is 10^{-27} grams!), they are so abundant that altogether they contribute about as much mass to the universe as all the stars combined. And, in addition, nobody knows what the mirror image of a neutrino looks like, which implies that neutrinos are like ghosts, alien space ships, and vampires.[1] Neutrinos march to a different drummer, and they seem to ignore the ordinary laws of physics. It fits the characteristics of this particle that the man who brought the neutrino into being was the Austrian physicist Wolfgang Pauli.

If Werner Heisenberg was (literally) the fair-haired boy among Arnold Sommerfeld's students in Munich, then his friend and fellow student Wolfgang Pauli (Fig. 3.1) was his dark counterweight.[2] While the blond Heisenberg went hiking in the Alps or swimming in Bavarian lakes, the dark-haired Pauli wandered restlessly through the bars and cafes, drinking excessively, then working afterward throughout the night. In the morning he would miss his lectures, and while a treatment by the famous psychoanalyst C. G. Jung didn't relieve his problems, it initiated a remarkable correspondence.[3] While Heisenberg loved to speculate, Pauli was feared for his perfectionism and skepticism. His colleagues even nicknamed him the "conscience of physics"; he judged poor work as totally wrong or—even worse—as "not even wrong." And on one particular December evening in 1930, Pauli had much better plans than to take part in a physics conference in the German university town of Tübingen: He wanted to attend a ball in Zurich that was "not to be missed." At least he apologized to his colleagues, in a letter that is today legendary, addressing them as "Dear Radioactive Ladies and Gentlemen."[4]

And Pauli, who liked to make his results public in letters rather than in journal articles—"I can afford not to get cited"[5]—included a scientific firecracker in the letter. At the top of the agenda of the meeting in Tübingen was the problem of energy conservation in radioactive beta decay. What was extremely puzzling in this process was that the energy of the decay products didn't add up to the energy of the original nucleus. It seemed that energy was lost—in contradiction to one of the most sacred laws of physics, the law of energy conservation. In his letter, Pauli couldn't resist offering a solution to the problem. As a "desperate remedy," he offered a really outrageous proposal: A new, electrically neutral particle that had not been discovered so far could carry the energy away secretly. The neutrino was born. (That name came later; Pauli called it a neutron.)

Soon after Pauli's neutrino postulate, the Rome group around Enrico Fermi and his "Via Panisperna boys" assumed the lead in figuring out the puzzling properties of the new particle. Fermi was to become a Nobel Prize winner in 1938, and he later led the construction of the first nuclear fission reactor in Chicago, then becoming a major contributor to the US atomic bomb project, the direct competitor of the German *Uranverein*. When he was only twenty-one years old, Fermi had earned his PhD in Pisa and afterward became— just as Heisenberg did a year later—a postdoc with Max Born's group in Göttingen. In 1933 he worked out a theory of nuclear beta decay, which combined Pauli's neutral-particle hypothesis with Dirac's ideas on the creation and annihilation of antiparticles and Heisenberg's description of proton and neutron as the two possible incarnations of the nucleon. To Pauli's particle he gave the name *neutrino* (Italian for "little neutral one")—to differentiate it from the heavier nuclear component, the neutron. More important, however, was that Fermi had developed a theory of neutrino interactions, and thus, at least in principle, a way to see the so-far purely hypothetical particle in an actual experiment. Despite the fact that in the following years more and more properties of the puzzling particle were reconstructed from theory, the neutrino itself remained as elusive as a ghost. Pauli himself was worried about this problem, writing soon after his neutrino hypothesis: "Today I did an awful thing, something no physicist

should ever do. I proposed a particle which never can be experimentally verified."[6]

This situation remained unchanged until, in 1951, two physicists from the US nuclear weapons lab at Los Alamos, Fred
Reines and Clyde Cowan, met by chance at the Kansas City
airport. Reines was thirty-three years old, and he was in a
complicated phase of his life. The US fission bomb project was
more or less completed, many scientists had left Los Alamos,
and Reines found himself staring at a blank wall, trying to
identify a problem that could become his life's work. He and
Cowan agreed that the discovery of the neutrino would be a
first-rate challenge.[7] The two physicists thereby embraced the
"think big—can do" philosophy that prevailed during World
War II in Los Alamos: If a task seemed interesting enough, it
could somehow be conquered. Reines was used to working on
bomb tests where one would trench out half a Pacific island
and pile it on the other half—just to blow up everything in a
gigantic explosion. And in every nuclear explosion, not only
huge amounts of energy were set free, but also billions upon
billions of antineutrinos. If these antineutrinos hit an atomic
nucleus, they would transform protons into neutrons and positrons, the antiparticles of electrons (Fig. 7.1).

Cowan and Reines developed the wild idea of utilizing a
nuclear explosion to detect these peculiar particles. The first
approach they outlined was to construct a remote-control
detector just forty meters away from a nuclear explosive device, which would record the annihilation radiation of the
created positrons when they annihilate with electrons. To keep
the detector from getting ripped apart by the shock wave of
the explosion, Cowan and Reines imagined a fifty-meter-deep

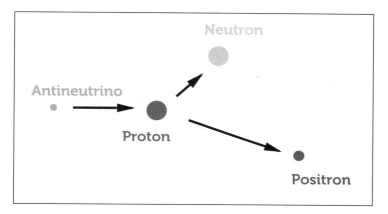

Figure 7.1. Detection of antineutrinos via inverse beta decay. An antineutrino transforms a proton into a neutron and a positron, the antiparticle of the electron.

evacuated funnel through which the detector was supposed to fall freely and finally land softly on a bed of feathers and foam rubber. The plan was to return to the test area a few days after the surface radioactivity had sufficiently diminished, dig out the detector, and "learn the truth about neutrinos."[8] Amazingly this bizarre enterprise was actually approved by the Los Alamos lab authorities, even though the measurement time was estimated to be too short to detect the theoretically predicted, extremely weak neutrino interactions. One evening in the fall of 1952, while a detector named El Monstro was being developed and the funnel was already being excavated, J. M. Kellogg, head of the Physics Division at Los Alamos, asked Reines to reconsider the possibility of using neutrinos from a nuclear reactor instead of from a bomb. The point was that in spite of neutrino fluxes from nuclear reactors being a factor of a thousand smaller than those set free in a nuclear explosion, a nuclear reactor could take data for a much longer period than the two seconds a detector could record a bomb detonation. Another idea of how to improve the measurement of the neutrons created in the neutrino detection and thus to better achieve discrimination from the background due to neutrons, as well as from gamma radiation produced in the reactor, made the reactor idea a serious option. In the spring of the following year the detector Herr Auge, for Project Poltergeist (Fig. 7.2), was ready for action. Later that same year Cowan and Reines reported the first evidence for neutrino reactions. It would, however, take three more years, several detector improvements, and a change of the reactor site before they finally could telegraph Pauli that—beyond any doubt—the neutrino had been discovered. Pauli, who at the time was staying at the European center for particle research, CERN, interrupted the meeting he was attending to read out the telegram to all participants. Then he gathered some friends to empty a bottle of champagne. Reines didn't get any of the champagne, but received the Nobel Prize in 1995.

The theory of neutrino interactions would not be the last contribution of Fermi's group to neutrino physics, in part because Fermi was not only one of the greatest researchers of the twentieth century, he was also an exceptionally gifted teacher: Seven of his students received Nobel Prizes themselves! In the 1930s in Rome, Fermi had started to gather a group of extremely talented students around him. In the end, Fermi was not only the godfather of the new particle; in addition, he trained two physicists who would shape the course of neutrino physics in the years to come (Fig. 7.3).

Figure 7.2. Clyde Cowan (left) and Fred Reines (right) with their working group and the detector *Herr Auge* (German for "Mr. Eye"). (Photo courtesy Los Alamos National Laboratory)

The first of these students to realize that the new member of the particle zoo was special was Fermi's genius student Ettore Majorana. Majorana was a Latin-looking Sicilian with dark eyes and a tragic expression on his face.[9] He didn't fit at all into Fermi's lively bunch of students. Like many great physicists, he was rather withdrawn and shy, and his mood would alternate from coyness to self-doubt to dismissive arrogance. In addition to that, he was agonizing over a brutal crime his uncle was accused of, probably wrongfully—a baby

Figure 7.3. Fermi's genius students, whose efforts were vital for the fate of the neutrino: Ettore Majorana (left), who discovered that neutrinos could be their own antiparticles and who disappeared without a trace in 1938; and Bruno Pontecorvo (right), who discovered the phenomenon of neutrino oscillations. (Left: courtesy AIP Emilio Segre Visual Archives, E. Recami and Fabio Majorana Collection; right: courtesy AIP Emilio Segre Visual Archives, Physics Today Collection)

was burned in its cradle. If Pauli was the conscience of physics, then Majorana—who was no less pedantic and no less a perfectionist—was known as the great inquisitor. Majorana had simply shown up in Fermi's institute one day in 1928 with a plan to change his course of studies from engineering to physics. On that occasion, Fermi explained to him his recent work and allowed him to glance at some results on his table. Majorana went home and spent the night recalculating Fermi's results with a self-developed algorithm. The next day Majorana went back to Fermi and asked if he could have another glance at the table. After he had convinced himself that his results and Fermi's were the same, he generously assured the bewildered Fermi that his results were correct. Afterward, Majorana, whose genius was compared by Fermi to that of Isaac Newton and Galileo Galilei, became a loose member of Fermi's group, one who appeared, however, only irregularly at the institute and largely ignored the conventions of the scientific community, even showing a certain contempt. For one thing, he refused to publish most of his results. Typically Majorana would take the tram in the morning to get to the institute and, during these trips, scribble his ideas and calculations on a pack of cigarettes. As soon as he had finished the pack he would throw it away. In this way, supposedly, his theory that the atomic nucleus was made out of protons and neutrons got discarded months before James Chadwick and Heisenberg arrived at the same insight. He even forbade Fermi to mention his results in a physics conference in Paris unless—according to his ludicrous condition—Fermi would incorrectly credit a colleague for this discovery, a colleague who would also be present at the meeting and whom Majorana considered to be totally inept. In 1933 Majorana visited Heisenberg for a couple of months in Leipzig. After this encounter, which the Italian author Leonardo Sciascia considered to be most important for Majorana, he started to appear increasingly gloomy and even misanthropic. He withdrew even further and stopped cutting his hair: "Nervous fatigue" was the diagnosis of his doctors. On March 26, 1938, Majorana finally disappeared, traceless, like the seemingly unobservable neutrinos, most probably from the Naples-to-Palermo ferry. He had bought a ticket for this vessel, at any rate, and

the ticket had been used. The circumstance of Majorana's disappearance was highly disconcerting to the bureaucracy of
Fascist Italy. Even Mussolini himself allegedly issued an order
when he was informed about the case: "I want him to be
found!"[10]

Nevertheless, Majorana remained lost, and his disappearance remains mysterious even now: Why did Majorana write
two letters full of ambiguities the previous evening, in which
he announced his disappearance and then revoked that an-
nouncement? Why did he ask his family not to wear sables?
Who was the man who slept in Majorana's bunk on the ferry
and who left the ship on arrival in Palermo? Why did Majorana take his passport and all his recently withdrawn money
with him? Who was the person that Majorana's nurse and a
Neapolitan monk identified as Majorana long after his disappearance? And who was the "great scientist" who allegedly
later lived in the monastery San Giovanni degli Eremiti in
Palermo?

There remains no end of uncertainty and speculation. Was
Majorana fleeing the responsibilities that came with the professorship he had grudgingly assumed the previous year at
the University of Naples? Did he anticipate the horrible de-
structive power of nuclear research to which he—as Sciascia
speculated—didn't want to contribute? Was he, as assumed
in contemporary conspiracy theories, kidnapped by a hostile
power? Or, tortured by depression and loneliness, did he com-
mit suicide—a "victim of science," in his combination of genius and mania, as his mother had feared?[11] In 2011 it made
news that a photograph had emerged that displayed a convincing similarity to Majorana, taken at an apparent sighting

of him in Venezuela in 1955.[12] Beyond any doubt the most inventive and bizarre interpretation of Majorana's disappearance was proposed in 2006 by the Ukrainian cosmologist Oleg Zaslavskii:[13] Majorana's disappearance was nothing but a *quantum hoax*. With his two contradictory letters, which Majorana wanted to make sure arrived at the same time, Majorana sought to illustrate the living-dead superposition of Schrödinger's cat using his own person.

The only sure thing in this story is that Majorana was responsible for a crucial breakthrough in the comprehension of the neutrino. In his last work, the "Theory of the Symmetry of Electrons and Positrons," Majorana realized that the neutrino, in contrast to all other known fermions, can be *its own antiparticle*. This is because antiparticles, which were proposed by Dirac, differ from particles in particular by having the opposite electric charge. The neutrino, however, is the only fermion that is electrically neutral—chargeless—and thus can well be identical with its antiparticle. The number of leptons minus the number of antileptons—*lepton number*—wouldn't remain constant in this case. Or as particle physicists put it, lepton number conservation is violated. Whether the neutrino is really a so-called *Majorana particle,* and not, like its charged brothers, a *Dirac particle,* is so far not settled. But it is exactly this property that most particle physicists assume determines the specific role of the neutrino. It could, for example, make the neutrino responsible for the fact that the universe consists of matter with barely any antimatter, or explain why neutrinos could convey information about a Grand Unified Theory of all particles and forces. The questions raised by Majorana remained moot for many years because the difference

between Majorana particles and Dirac particles is only relevant for particles possessing mass. This is due to the fact
that—as we saw in Chapter 4—only left-handed particles and
right-handed antiparticles can participate in weak interactions. For the neutrino, that implies that in order to decide
whether a particle is its own antiparticle, one has to transform right-handed antineutrinos, which are produced, for
example, in nuclear beta decay, into left-handed neutrinos.
This is only possible, however—due to the *chirality flip* involved—if neutrinos possess mass. As it happened, the question of whether neutrinos actually possess mass was only an-
swered some sixty years after Majorana's last work and
disappearance.

In the years following its discovery, the neutrino mutated
into an efficient probe for new phenomena and a cash cow for
Nobel Prizes. Only half a year later Chien-Shiung Wu (or Ma-
dame Wu, as she was most often known) discovered that weak
interactions affect only left-handed particles and that neutrinos, which, as electrically neutral particles, possess only weak
interactions, must always be left-handed. All neutrinos that
can be detected with the known interactions are thus left-
handed and all antineutrinos right-handed. Whether a mirror image of the neutrino—a right-handed neutrino—exists
at all, is not yet settled. In 1963, Leon Lederman, Melvin
Schwartz, and Jack Steinberger (who were honored with the
Nobel Prize in 1988) discovered the muon neutrino, and with
it the first hint of the second particle family. And in 1987, the
huge underground detectors IMB in America and Kamiokande in Japan (Nobel Prize 2002) counted nineteen neutrino
events within twelve seconds, originating from the Supernova

1987A, a giant star explosion in the Large Magellanic Cloud, 150,000 light-years away. (The Baksan detector in Russia added another five to this count.) At the same time, Ray Davis was collaborating with John Bahcall to detect neutrinos from the sun. And he opened wide a door into a new chapter in the history of neutrino physics, the search for neutrino oscillations—a phenomenon that, in its original form, was predicted in 1957 by Fermi's student Bruno Pontecorvo.

Just as Heisenberg was the counterweight to Pauli, Pontecorvo could be considered the counterweight to Majorana. When only eighteen years old, Pontecorvo, the son of a wealthy Jewish-Italian family, whose brothers became a geneticist and a movie director, started to attend Fermi's lectures. In the years that followed, Pontecorvo became one of Fermi's closest assistants. After a research visit with the Joliot-Curies in Paris, where he fell in love and became the father of a baby, he could not return to his Italian home country because of the anti-Semitic policy of the Fascist regime. A long odyssey began. When German troops occupied Paris, Pontecorvo fled first via Spain to the United States, where he temporarily worked for an oil company. Then he joined a lab in Canada, where he was supposed to contribute to the British nuclear bomb project. Finally, in 1950, he was offered a professorship at the University of Liverpool, which he occupied until late summer of the same year.

On August 31, 1950, about twenty-two years after Majorana's disappearance, Pontecorvo disappeared as well. And if one can interpret Majorana's disappearance as a quantum hoax, then one might want to interpret Pontecorvo's disappearance—he reappeared five years later in the Soviet Union—

as a hoax anticipating neutrino oscillation, thereby illustrating his greatest discovery. Apparently Pontecorvo, his wife, and his three sons left their Italian holiday retreat and travelled via Sweden to Finland, where they boarded a Soviet submarine and emigrated to Russia.[14] Pontecorvo's emigration stirred up some Cold War excitement in the Western world, happening, as it did, in the same year in which the physicist Klaus Fuchs, who had worked on both the UK and the US nuclear bomb projects, had been uncovered as a Soviet spy. Pontecorvo himself, however, was never convicted nor even suspected of espionage. In Russia he joined the Joint Institute for Nuclear Research in the science city of Dubna, north of Moscow on the banks of the Volga, where he concentrated on fundamental research in particle physics and in particular neutrino physics.

Here Pontecorvo discovered that neutrinos, if they actually possess mass, would have an absolutely stunning property: The three different neutrino families, or flavors—electron, muon, and tau neutrinos—could transmute into each other. More accurately, an electron neutrino, for example, would oscillate back and forth among these three identities, until a measurement collapsed (or decohered) its quantum mechanical wave function (Fig. 7.4). The whole mechanism is another example of the weird properties of the quantum world. Just as Schrödinger's cat can be dead and alive at the same time, one cannot assign a single individual mass to an electron neutrino. Rather electron, muon, and tau neutrinos have to be understood as superpositions of all three masses. What physicists call *mixing* determines how strongly the individual masses contribute to a flavor state. An electron neutrino pro-

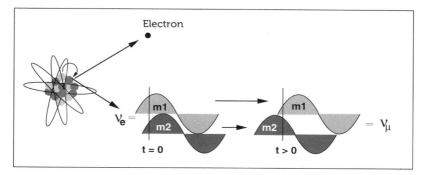

Figure 7.4. Neutrino oscillations. In the radioactive beta decay of an atomic nucleus a neutron gets transformed into a proton. Thereby an electron and an electron-(anti)-neutrino are produced. The neutrino propagates as a superposition of two waves that correspond to the two masses m_1 and m_2. Assuming that the electron neutrino at time t_0 consists of 60% m_1 and 40% m_2, at a later time $t > t_0$ the faster wave of m_1 has outrun the wave of m_2. The new superposition corresponds to 60 percent m_2 and 40 percent m_1, and thus to (the assumed composition of) a muon-(anti)-neutrino.

duced in a nuclear reaction propagates as a superposition of all three masses through space, and the different masses advance with different speeds: the less the mass, the less of the total neutrino energy is being used up by it, and the more is available as kinetic energy (energy of motion). But if now the different masses propagate with different speeds, the composition of the total quantum mechanical wave functions changes. Consider, for instance, a neutrino that can be described at its production point by a wave crest of the first mass and a small amplitude of the wave function of the second mass—the superposition corresponding to an electron neutrino. Soon the wave trains of different speed will be misaligned. Now the contribution of the second mass will be greater, and this superposition corresponds to a muon neutrino. Consequently a neutrino being created as a specific flavor (electron, muon, or tau) at one point can be observed somewhere else with a well-defined probability as a different flavor. This phenomenon is called neutrino oscillation, and it soon turned out to be the best option for probing neutrino masses. Since detecting neutrinos at all is a huge challenge, finding neutrino oscillation is an enormous task. In a thrilling pursuit spanning decades, ever more gigantic tanks of water and other liquids were placed in laboratories thousands of meters underground—a real tour de force. Finally, in the 1990s, the pursuit culminated in a series of spectacular discoveries. The golden age of neutrinos had begun.

Nuclear Decays a Thousand Meters Underground

In 1994 I was thrown into a world of madmen, dreamers, and visionaries, a heterogeneous bunch of nuclear and particle physicists, astrophysicists, chemists, and engineers who, far away from their original areas of expertise, were setting off on a quest for the neutrino. In the fall of that year I had passed my diploma exam (the German equivalent of the MSc) more poorly than well and was now looking for an interesting research topic for my dissertation. I knew more or less what I wanted: I was fascinated by the possibility of uncovering symmetries in the fundamental laws of nature. And I wanted to work on theoretical physics, because I was more interested in the beauty of the laws of physics than in the machines employed to wrest these laws from nature. After I couldn't find an inspiring research topic at Heidelberg University's Institute of Theoretical Physics, I recalled something I had read when I was still studying for my exam, about a specific nuclear decay that could shed light on the masses of neutrinos.

Neutrino masses that do not occur in the Standard Model are predicted in GUT theories and could yield information about the most fundamental facts of nature. This was a topic that I found enthralling even back in high school. So now, in my search for a dissertation topic, I replied to an advertisement by a research group at the Max Planck Institute for Nuclear Physics (MPIK), looking for students to work on the theoretical background of such an experiment. MPIK—technically not a part of the university—was placed by its founders on a forested hill just outside Heidelberg, in order to pacify the city population's fear of hazards related to nuclear research.

When I first drove up that hill, the vast majority of particle physicists still assumed that neutrinos were massless. But there was noticeable unrest—stimulated mainly by GUT theorists whose models predicted neutrino masses—by astrophysicists who were looking for a suitable candidate for the dark matter in the universe, and by nuclear physicists who needed neutrino masses to understand the hints of neutrino oscillations they were finding in their underground experiments. Around this time, a poll initiated by the German Ministry for Research, looking for the most vexing problems in nuclear and particle physics, ranked as number one the question about the existence of neutrino masses.

My prospective PhD advisor at MPIK, Hans Volker Klapdor-Kleingrothaus, had, in a startling coup, gotten hold of almost eighteen kilograms of nearly pure germanium 76 from centrifuges of the Soviet nuclear weapons program. But he didn't want to build weapons or reactors; all he had in mind was to learn about neutrino mass by observing something called neutrinoless double beta decay, in which two electrons and no neutrinos are emitted. And, unlike the weapons-capable materials produced in comparable centrifuges, this material wasn't highly radioactive. On the contrary: It would take a hundred billion times the age of the universe before even half of the germanium nuclei would have decayed.

A decay occurring so rarely is extremely difficult to observe. The major problem here is that the natural radioactivity caused by radioactive isotopes in the materials of the environment is many times stronger than the radiation of the double beta decay itself. Without taking extreme measures to suppress it, this "background" would bury the measurement signals under piles of natural radiation. So the rare, exciting observation one is looking for has to be filtered out from uninteresting signals. It is indeed like looking for a needle in a haystack. Several strategies exist to solve this problem: One can improve the search itself; physicists do this by developing more and more advanced computer programs for their data analysis. Or one can shrink the haystack. The radioactivity of the environment is being renewed continually by the bombardment of cosmic radiation—very energetic particles that streak in from the depths of the universe and can transform stable nuclei into unstable ones. The higher the altitude the greater is this effect, since less protective atmosphere means less screening against the bombardment. So if you want to keep the precious germanium powder as pure as possible, you better not ship it by plane.

Herbert Strecker, the technician of the group, had the delicate job of carrying a few suitcases with white powder worth millions of dollars through the Eastern Bloc and across the Iron Curtain from Russia to Heidelberg. One can easily imagine the lively discussions Herbert must have had with customs officers at various borders: "Is this cocaine?" "Oh no, this is just highly enriched nuclear material"!

From Heidelberg the material was shipped on a container vessel—always stored as deep in the hull as possible below the waterline—to the United States, where it was converted into crystals usable in the experiment that was to be situated in the tunnels of the Italian Gran Sasso Laboratory. The experiment itself is a nice example of how sophisticated laboratory equipment has to be in order to reduce the background haystack and attempt to see through the neutrino's game.

Whoever wanted to visit the Heidelberg-Moscow experiment first had to drive into the ten-kilometer-long highway tunnel connecting the Adriatic coast with Rome (see Fig. 8.1). The tunnel extends under Europe's southernmost glacier in the Abruzzo mountains, on a peak of which Mussolini was imprisoned before he was freed by German paratroopers. During the drilling of the tunnel, seven people died when an aquifer was encountered. Half way through the tunnel, now 1,400 meters below blue sky and snowy peaks with grazing sheep and sleepy villages where Pecorino cheese is manufactured, you take a turn and come to a stop in front of a large metal gate. You feel just like James Bond when you say the password into the intercom, the gate opens slowly with an orange warning light flashing, and armed security personnel ask for your ID card. Then you enter the largest underground lab in the world, made possible only by the unflagging commitment of Italy's particle godfather, Antonino Zichichi, descendent of a 700-year-old Sicilian family and on friendly

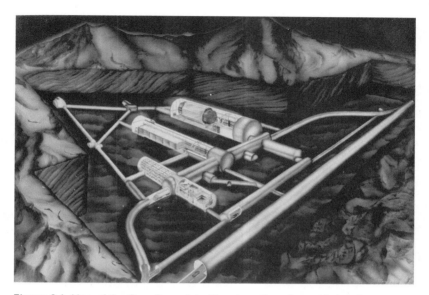

Figure 8.1. Map of the Gran Sasso lab. (Courtesy Gran Sasso National Laboratory)

terms with the pope and the Italian Christian Democratic party and, according to rumors, even having admirers in the Mafia. There are three huge halls, each a hundred meters long, eighteen meters high, and twenty meters wide, connected by long corridors. It is cold and dark, and water drips from the rock walls. In almost twenty different experiments, separately enclosed, researchers here look for neutrinos, rare nuclear decays, and dark matter. The reason for this whole endeavor being underground is, again, the cosmic radiation: While at the earth's surface 400,000 muons per hour and per square meter volley down on you, inside the Gran Sasso lab, shielded by 1,400 meters of rock, there is only one muon per hour per square meter left.

In comparison with some of the other experiments—for example the MACRO experiment searching for magnetic monopoles and for supernova neutrinos from the explosions of burned out stars, with its enormous dimensions of seventy-six by nine by twelve meters—the Heidelberg-Moscow experiment was not very impressive on first sight. The detectors had the approximate size of beer cans and were hidden in boxes about a cubic meter in size. But as boring as the experiment seemed to be from the outside, the details of the setup were exceedingly sophisticated: In order to get rid of the effects of even the one muon hitting the detector every hour or so, the boxes were covered with detectors of a special kind, whose function was to indicate when a muon penetrated the setup so that the data from that moment could be discarded. The source material, as I've already described, was ultra-pure germanium, as free of contamination as anything ever manufactured. The purpose of its purity was to assure that it contained almost no radioactive nuclei of other elements that could trigger the detector with confusing signals. To avoid contamination from the detector material, the scientists used another trick. Just as the ice-cream cups at McDonald's can be eaten along with the ice cream to avoid unnecessary waste, the detectors were identical to the source, meaning that they would detect their own decay. Then the detector was surrounded by lead shielding in order to screen out the radioactivity of the surrounding material—and not normal but ultra-pure lead. (Some competing experiments even used lead from the keels of ancient Roman galleys that had lain for more than 2,000 years on the bottom of the sea, largely unaffected by the activating effect of cosmic radiation.) Finally, each experimental box was connected to several nitrogen gas bottles that "rinsed" the whole setup and thus prevented radioactive gas particles that escaped from the tunnel's walls from migrating into the detector box. As you can well imagine, the entire procedure yielded a massive reduction of environmental radioactivity: Only 0.2 events per year and per kilogram of detector material were eventually observed in the energy region of interest in the super-sensitive device. In fact, the detectors were so sensitive that as a by-product they could even indicate if they were penetrated by dark-matter particles which, according to astrophysicists, are supposed to populate the universe. These particles could hit atomic nuclei inside the detector, causing them to recoil and rip loose some electrons in the atom, which would then be recorded by the detector. This allowed the Heidelberg-Moscow experiment to also provide, apart from its quest for the neutrino mass, what was at that time the most sensitive limit on the existence of dark-matter particles.

Klapdor-Kleingrothaus was understandably proud that his experiment—the Heidelberg-Moscow search for *neutrinoless double beta decay*—was the most sensitive one in the world, and that it had the best chance to reveal the neutrino mass. This was true, however, only if the neutrino is a Majorana particle, as I shall now explain.

In general, a double beta decay is nothing but two radioactive beta decays occurring simultaneously in one nucleus. During ordinary ("single") beta decay, a neutron is transformed into a proton, an electron—the "beta radiation"—and an anti-neutrino. Consequently, in a double beta decay, two neutrons are turned into two protons and two electrons. Typically also two antineutrinos are emitted, but possibly not in all cases. For if the neutrino is a Majorana particle, meaning identical to its own antiparticle, and if it has a mass, the following decay process is also possible: A neutron decays into a proton, an electron and—as dictated by the weak interaction—a right-handed antineutrino. Being a Majorana particle, the right-handed antineutrino is the same as a right-handed neutrino, and via its mass (see Chapter 6) it can be metamorphosed into a left-handed neutrino. This left-handed neutrino can be gulped by another neutron in the nucleus and in this way trigger the second decay (see Fig. 8.2). Consequently the entire decay proceeds without the emission of any neutrino, and the bigger the neutrino mass the more likely is such an event. And since no antineutrino or neutrino can carry any energy away, the entire decay energy flows into the two emitted electrons that are detected in the experiment. The experiment then yields a pileup of signals, each with the total decay energy; such a sharp peak at the endpoint of the energy spectrum is the smoking gun the experimenters are looking for.

But what if the neutrino is *not* a Majorana particle? In that case, one possible way to extract information about neutrino masses is to look for the cosmological consequences of massive neutrinos (see Chapter 10). Or one can go back to Pauli's original motivation for proposing the neutrino in the first place: the missing energy in the nuclear beta decay. According to Pauli, this energy is carried away by the emitted neutral particle. If the neutrino, as we now call it, is massless, this energy can be arbitrarily small. But if the neutrino possesses a mass, the emitted neutrino has a minimum energy—just the energy mc^2 of its mass, which it possesses even if it is not

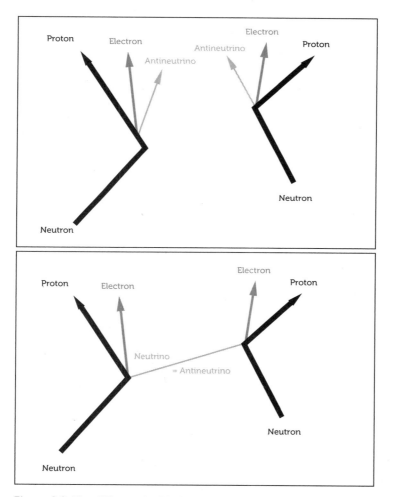

Figure 8.2. The different double beta decay processes. Upper picture: the decay mode allowed in the Standard Model with the emission of two antineutrinos. Lower picture: the neutrinoless decay where a massive Majorana neutrino is being exchanged between the two decaying neutrons.

moving. In that case, it is not possible for the entire decay energy to be supplied to the emitted electron. As a consequence the energy spectrum of the electrons should not terminate exactly at the total decay energy but somewhat below it. The corresponding kink in the spectrum could be observed, and if one can determine the electron energy with sufficient accuracy, it provides information about the mass of the neutrino—whether it is a Majorana particle or not.

In order to reach the necessary sensitivity, however, both a proper source and a powerful spectrometer are necessary. A good option for the source is the heavy hydrogen isotope tritium, which decays sufficiently rapidly (its half-life is 12.3 years) and whose decay energy is so small that if the neutrino has mass, the spectrum gets distorted significantly. The needed spectrometer is a device that uses magnetic and electric fields to measure a charged particle's energy; first a magnetic field is used to collect and guide the electrons emitted from the source into the spectrometer and deflect them in a way that they run almost parallel to the magnetic field. The resulting parallel beam of electrons then runs against the repulsion of an opposed poled electric field. As electrons with smaller energies get reflected by the electric field and don't reach the detector by varying the electric field, the electron energy can be determined. Searches for neutrino mass using tritium had been conducted over the years, all without finding any sure evidence of a non-zero mass. (The sensitivity of these earlier experiments was, in fact, less than was available in the more recent double beta decay experiments and from cosmology; see Chapter 10.)

To search for a very small neutrino mass using tritium decay, a much-improved spectrometer was needed. Such a machine was built between 2004 and 2006 for the KATRIN experiment in Karlsruhe, and did indeed break all records. It was ten meters high, ten meters wide, and twenty-four meters long. Within its stainless-steel hulk it contained the biggest vacuum cavity in the world. Not surprisingly, it also broke the size restrictions set by the height of German highway bridges. This meant that there was no way to transport the device by road from where it was manufactured in Bavaria to where it was needed in Karlsruhe. Instead of travelling 400 kilometers on German highways, the 200-ton colossus went on a 9,000-kilometer odyssey by boat, first down the Danube river into the Black Sea, then on three different vessels through the Bosporus, Adriatic Sea, Mediterranean Sea, and Atlantic Ocean, rounding the whole of Europe to reach the mouth of the Rhine, from where it went upstream to the small village of Leopoldshafen. It was finally loaded, with Europe's biggest heavy-duty crane, onto a fourteen-axle flatbed trailer, which crawled at two to three kilometers per hour to the lab at the Karlsruhe Institute of Technology (KIT). Traffic islands, traffic lights, and street lighting, as well as the trolley wires of two tram lines had to be taken down to make way for the detector. Even then, only a few centimeters of space were left for the gigantic device to maneuver around obstacles (see Fig. 8.3). After several years of measurement, KATRIN is supposed to reach a neutrino mass sensitivity of 0.2 eV—a number that is interesting both in view of cosmological implications and in view of a controversial result of the double beta decay search.

Figure 8.3. Like a spaceship between the houses: transport of the gigantic KATRIN spectrometer. The village of Leopoldshafen is in a state of emergency. (Courtesy Karlsruhe Institute of Technology)

In the spring of 2004, four years after I had left the research group at MPIK Heidelberg and at about the time I received my job offer from Hawai'i, Klapdor-Kleingrothaus was convinced that he had found the signal for neutrinoless double beta decay in his data records. The particle physics community remained skeptical, though. This was partly because the new signal was found in a largely old data set from which, in a previous analysis, a signal of this strength had been excluded. So the discovery was due not so much to new data as to a new method of analyzing old data. Moreover, there existed several pileups of signal events—so-called *peaks*—in the interesting part of the energy spectrum, not only the one at maximum decay energy. While the other peaks remain unexplained and thus suggest that unknown radioactive background exists, the one at the total decay energy could be explained by neutrinoless double beta decay, but not definitively so. It could also be produced in the same way as the other peaks, by the radioactive background. Moreover, at this time members of the collaboration were quarreling. The Russian members didn't agree with Klapdor's interpretation, and also the authors of the article announcing the discovery didn't even agree among themselves. Klapdor-Kleingrothaus himself maintains that the change in the analysis was justified and natural because better statistics allowed a better fit to background lines. He also points to independent analyses that argue against an interpretation of the signal peak as background. Meanwhile first results of the EXO-200 experiment in New Mexico—performed by, among others, Stanford University's Giorgio Gratta and Klapdor-Kleingrothaus's former PhD student at the University of Alabama, Andreas Piepke—were published in May 2012. These results do not confirm the discovery claim but also don't unambiguously rule it out. As of this writing, the situation is still unclear. There certainly are good reasons to be skeptical, while, on the other hand, no experiment is sensitive enough yet to refute Klapdor's claim.

Major competitors in the race to confirm or disprove Klapdor-Kleingrothaus's claim are EXO-200 in the United States, KamLAND-Zen in Japan, and CUORE in Italy, as well as the GERDA experiment in Italy, which started to take data in 2011 and will operate for some five to ten years. GERDA was inspired by the GENIUS proposal, which rests on an idea of the MPIK underground expert Gerhard Heusser and was developed in Klapdor-Kleingrothaus's group by Jochen Hellmig, Laura Baudis, and Bela Majorovits. The basic idea of their proposal is that the major part of the background radiation does not originate in the detectors themselves but in their environment. Since the detectors have to be cooled down in order to operate at all, the GENIUS proposal advocated the bold idea of cooling the detectors not through some separate cooling system, but by hanging them directly inside the cooling liquid. Either liquid nitrogen, as in the GENIUS proposal, or liquid argon, as in the GERDA realization, can then act simultaneously as cooling device and as shielding against external background sources. Within a few years of measurement, GERDA will probably be able to test Klapdor-Kleingrothaus's claim of discovery.

In parallel with these developments, Klapdor-Kleingrothaus initiated a lively research activity in theoretical physics within his group, where I, as a budding theorist, could make some contributions and gain some valuable experience. A major topic that required calculations was how contributions of new physics might reveal themselves in neutrinoless double beta decay. After all, not just Majorana neutrinos but also every other new species of particle could be exchanged between the two decaying neutrons and in this way trigger the decay, so long as lepton number is changed by two units, just as it is for Majorana neutrinos. Actually Jose Valle, now at the University of Valencia, Spain, and his PhD advisor Joel Schechter of Syracuse University had shown in 1982 that neutrinoless double beta decay always implies a non-vanishing neutrino mass. Their theorem relies on the fact that in this case a neutrino can always fluctuate into its right-handed antiparticle by means of a neutrinoless double beta decay. The mass resulting from this fluctuation, however, can be tiny and need not constitute the dominant contribution to the decay.

So Martin Hirsch, the postdoc of our theoretical group and, for me, an inexhaustible source of knowledge, and Sergey Kovalenko of the Russian nuclear research center JINR, located in the once secret city of Dubna, had a feast analyzing various contributions to the decay: contributions from R-parity violating supersymmetry, originally proposed by Rabi Mohapatra and Kaladij Babu; contributions from leptoquarks; and, with Orlando Panella of Istituto Nazionale di Fisica Nucleare in Perugia, Italy, contributions of composite neutrinos. Hirsch and Kovalenko also worked out that neutrinoless double beta decays are possible in SUSY models without R-parity violation, if the SUSY partner of the neutrino, the sneutrino, has itself a Majorana property. In 2006 Hirsch and Kovalenko, in collaboration with Ivan Schmidt, generalized the Schechter-Valle theorem with the consequence that the discovery of neutrinoless double beta decay would imply also the existence of arbitrary processes violating lepton number conservation, changing lepton number by two units, processes that might be accessible at colliders such as the LHC. It was a fantastic time to work on neutrino physics in this environment. In my dissertation, drawing on an idea of Kovalenko and in collaboration with Hirsch, I developed a general classification of all possible contributions to neutrinoless double beta decay. In two further pieces of work I collaborated with Klapdor-Kleingrothaus, Alexei Smirnov, and Tom Weiler to analyze the mutual relations of double beta decay, tritium beta decay, and cosmology.

In subsequent years, MPIK, the Max Planck Institute in Heidelberg where I worked, became one of the most exciting places in the world for a young physicist who wanted to learn about the most fundamental theories of nature. In the adjacent building—with an all-too-evident spirit of competition with our boss—resided Till Kirsten, who was the spokesman of the GALLEX experiment, which searched for oscillations of solar neutrinos. Till Kirsten had already been the first to detect the neutrino-emitting double beta decay—the Standard Model version of the decay his opponent Klapdor-Kleingrothaus was now after—in subterranean reservoirs containing a radioactive isotope of selenium, using chemistry technology. Now Kirsten was using his expertise in nuclear chemistry for solar neutrino detection. Together with his collaborators, he had constructed an experiment that used chemical elution to extract sparse atoms of germanium from gallium, the germanium having been created from gallium atoms by neutrino capture. This experiment was sensitive to the lower energy neutrinos emitted by the sun. Since the flux of these low-energy neutrinos does not depend sensitively on exact details of the solar model that is used, the researchers hoped that they could finally find out from the comparison of theory and experiment whether solar neutrinos are indeed oscillating.

This was the situation in February 1996, when members of our research group were hanging around together in the MPIK coffee room and Jochen Hellmig slammed a *Physical Review Letters* article onto our coffee table and proclaimed: "The coming year will be the year of neutrinos." In the article a research group called the LSND Collaboration, which was producing neutrinos in an accelerator located in the US nuclear weapons laboratory in Los Alamos, reported a hint of neutrino oscillations. But Jochen wasn't right. There wasn't a year of neutrinos about to come. Instead it was to be an entire decade of neutrinos.

~ 9 ~

New Physics Is Falling from the Skies

Beyond any doubt this experiment was a monster. A thousand meters below the rock of a holy mountain in Japan, more than ten thousand electronic eyes would peer into a tank as broad as a ballroom and as high as a ten-story building, filled with 50,000 tons of ultra-pure water (see Fig. 9.1). What the eyes were looking for were bluish flashes of light. These flashes occur when a neutrino from the sun or the atmosphere collides with an electron in one of the water molecules, accelerating it to a speed greater than the speed at which light travels in water, thus breaking the optical analog of the sound barrier.

As soon as the eyes record such a flash, they transform it into an electrical signal and send it for processing into five big containers of electronic gadgetry adjacent to the tank. After about two years of gathering data with this monster, Takaaki Kajita stood up to give his talk on the measurements of atmospheric neutrinos at the Neutrino '98 conference. When he had finished—when he had said the words, "the

Figure 9.1. The inside of the monster: the 11,200 electronic eyes of the Super-Kamiokande experiment in Japan. (Courtesy Kamioka Observatory, Institute of Cosmic Ray Research, University of Tokyo)

Super-Kamiokande experiment has evidence for a non-vanishing neutrino mass"[1]—his audience was stunned. As Berkeley physicist Hitoshi Murayama (who, by the way, is responsible for the idea that the whole universe could have been created by neutrinos) remembers this moving moment: "Uncharacteristically for a physics conference people gave the speaker a standing ovation. I stood up too. Having survived every experimental challenge since the late 1970s the Standard Model had finally fallen. The results showed that at the very least the theory is incomplete."[2]

One day later, US President Bill Clinton delivered a speech in front of graduates of the Massachusetts Institute of Technology (MIT) in Boston: "Just yesterday in Japan, physicists announced a discovery that tiny neutrinos have mass. Now, that may not mean much to most Americans, but it may change our most fundamental theories—from the nature of the smallest subatomic particles to how the universe itself works, and indeed how it expands."[3] The fall of the Standard Model was the end of a thirty-year-old search for confirmation of Pontecorvo's idea that neutrinos could oscillate back and forth among the three flavors of electron neutrino, muon neutrino, and tau neutrino—a fervent race for the discovery of neutrino mass.

When this story started, it was about neither atmospheric neutrinos nor a search for neutrino masses. It started with a scientist who wouldn't give up after losing one race and looked to the sun as a means of winning another. Ray Davis had competed with Reines and Cowan for the discovery of the neutrino and, as it turned out, he had bet on the wrong horse. While Reines and Cowan were detecting antineutrinos by their ability to turn protons into neutrons and positrons, Davis relied on an idea of Pontecorvo, namely to detect the antineutrinos produced in a nuclear reactor by having them transform nuclei of chlorine atoms into nuclei of the noble gas argon and electrons. What Davis didn't know was that this process is possible only for left-handed neutrinos and not for the right-handed antineutrinos that are produced in a nuclear reactor. And even if the neutrino were its own antiparticle, as proposed by Majorana, the transition from left-handed to right-handed would depend on the neutrino mass, and be very slow for a tiny mass. Although Davis experimented with buried tanks filled with several thousand liters of the cleaning and fire-extinguishing agent carbon tetrachloride and achieved a sensitivity twenty times better than Cowan and Reines, he didn't measure anything. But Davis, a bulldog and a maverick, wouldn't give in that easily. If chlorine wasn't able to detect reactor neutrinos, perhaps it would work with neutrinos from a different source? The sun seemed to be a promising possibility.

In his Caltech lectures Richard P. Feynman liked to tell his students the story of the British astrophysicist Arthur Eddington, who, according to legend, was sitting one evening with his girlfriend on a bench watching the night sky.[4] When his girlfriend exclaimed, "Look at the beautiful shining stars!" he simply replied, "Yes, and right now I am the only man in the world who knows why they are shining." What Eddington had correctly surmised was that nuclear fusion reactions proceed inside the sun and other stars, uniting lighter atomic nuclei into heavier ones and in this way producing the energy that is radiated as starlight. But these fusion reactions produce not only energy; they also spew countless neutrinos into space, billions of which bombard each square centimeter of earth each second. And, in contrast to the antineutrinos from nuclear fission processes in nuclear weapons or reactors, in stars we are really dealing with neutrinos and not with antineutrinos. This means that the same experiments with which Davis failed to detect reactor antineutrinos could now be used to detect solar neutrinos—particles produced 150 million kilometers away. In the end, it was the exploration of solar as well as atmospheric neutrinos that would shatter the cozy world of the Standard Model of particle physics.

At first, however, it didn't look as if Davis's luck would change. When he submitted his first paper containing an estimate of an upper bound for the solar neutrino flux, the referee wrote that the entire endeavor reminded him of an experimenter standing on a mountain and reaching for the moon, concluding "that the moon was more than eight feet from the top of the mountain."[5] Davis's task seemed more than Herculean.

In the early 1960s Davis found an ally, the former Louisiana tennis champion and later Princeton theorist John Bahcall, who had devoted his life to understanding the sun.[6] While Bahcall's first estimates of neutrino capture rates in chlorine did not make Davis's experiment look promising, it soon turned out that once contributions of higher-energy neutrinos that would bring the nucleus into an excited state were taken into account, the predicted rates of argon formation were almost twenty times as big. A year later Davis was able to confirm a crucial part of Bahcall's predictions experimentally. When he was phoned and told the result, Bahcall said later, it was the best moment of his entire career. The road to the detection of solar neutrinos seemed to be open, and Davis started to build a 400,000-liter tank to be filled with tetrachlorethylene, in America's biggest underground gold mine, in South Dakota.

But when the first new results came in, they were disappointing. In April 1968, Davis and collaborators published an upper bound on the solar neutrino flux that was a factor of seven below the theoretical predictions. Finally in 1970 the first neutrinos from the sun could be measured. This was an enormously important result, as it finally confirmed Eddington's idea that stars produce their energy by nuclear fusion. But the neutrino fluxes themselves remained puzzling: Even after Bahcall and collaborators had improved their calculations, a discrepancy of a factor of three remained. If Davis and Bahcall were right about this, this result could turn into a sensation: The missing electron neutrinos from the sun could be explained—corresponding to Pontecorvo's idea—with neutrinos oscillating into other flavors. At first the physics community was skeptical. Was it really possible that Davis measured the decay rates corresponding to neutrino capture this accurately? A fraction of a nuclear decay per day in a huge tank volume? And did Bahcall really understand the fusion reactions within the sun well enough to make such an accurate prediction? This was particularly questionable because the neutrinos captured in the chlorine experiment were not directly created in the fusion of protons into deuterons (nuclei of a heavy hydrogen isotope), which is the process responsible for the major part of the energy production in the sun and thus directly related to the sun's temperature.

Instead they originated from a subsequent reaction, whose rate was much more difficult to estimate. The only way out seemed to be an experiment capable of detecting the lower-energy neutrinos that came directly from the proton fusion. Such an experiment had been proposed as far back as 1965 by the Russian theorist Vadim Kuzmin, but it wasn't tackled seriously until the end of the 1980s. Yet even if Davis and Bahcall's results were confirmed, there still remained another problem: In order to explain the deficit of solar neutrinos in Davis's experiment by neutrino oscillations, the distance of earth and sun had to be "fine-tuned" exactly to correspond to the neutrino mass differences. This was an explanation that most physicists considered to be rather unlikely.

However, in parallel with the efforts of the experimenters and the nuclear physicists, the particle theorists had also made progress.

First, Lincoln Wolfenstein, a theorist at Carnegie Mellon University in Pittsburgh, Pennsylvania, had discovered in 1977 that the mixing of neutrinos gets altered if they propagate not in empty space but in matter. The reason is that neutrinos get decelerated by interactions with matter. From outside, this looks as if the neutrino had more mass; a smaller part of its total energy can appear as kinetic energy. Now, Wolfenstein is an extremely modest and unselfish person. As Palash Pal, now a scientist at the Saha Institute in Kolkata told me, Wolfenstein applauds any colleague who solves an interesting problem, even if he himself or one of his students (like Palash) had worked on the same problem, competing for a solution, and now would get no credit for that work (Palash later wrote about this experience in a humorous essay on the subjectiveness of joy and sorrow). To push or even to influence someone else goes so much against Wolfenstein's nature that when students ask him for a deadline for submitting their work, he tells them, after a long hum and haw, to set the deadline themselves.[7] So naturally he didn't consider his work on neutrinos to be anything more than a pastime, not worth looking into more deeply. Three years later, Vernon Barger, Kerry Whisnant (both at the University of Wisconsin, Madison, by then), Roger Phillips (Rutherford Appleton Laboratory, UK), and my later boss Sandip Pakvasa of the University of Hawai'i worked on the mixing of neutrino flavors within matter. They found out that specific combinations of neutrino energies and matter densities can amplify the mixing significantly: The matter effects can cancel the mass differences between single neutrinos, with the result that wave packets with equal momentum run next to each other with the same velocity and the mixing becomes maximal.

The real breakthrough, however, was provided in 1986 by two Russians. In the mid-1980s, Alexei Smirnov was a young researcher at the Institute for Nuclear Research of the Soviet Academy of Sciences in Moscow. One day his colleague Stanislav Mikheev walked into his office and proposed that they study the interactions of atmospheric neutrinos with matter. "It was an obvious thing to apply the approach also to solar neutrinos,"[8] Alexei told me years later when I met him at a workshop in Beijing, China. What Alexei and Stanislav did was to look at how a neutrino born in the sun's core evolves. They made the stunning discovery that the electron neutrino— which, because of its large effective mass, is created in the heaviest state in the solar core—loses mass as it propagates toward the sun's surface through matter of decreasing density, turning slowly into a maximal superposition with the muon and tau neutrinos. The particle remains in the heaviest state even when it mixes more and more strongly with muon and tau neutrinos, which, in vacuum, are heavier. Once the neutrino leaves the sun it is therefore in what is called a pure mass eigenstate consisting predominantly of the muon and tau flavors; it doesn't oscillate any more until it reaches the earth. On arrival it can then, with a certain probability, be detected as a muon or a tau neutrino corresponding to the flavor composition of the heavy state in vacuum.

As Alexei told me, he immediately realized how important their discovery was. But it was a long time before the two Russians could convince the scientific community. First they sent their results to Wolfenstein, but he didn't believe them. When they submitted their work to a Russian journal, it was rejected. Only on their second try were they successful, getting it published in the Italian journal *Nuovo Cimento*. Next Alexei wanted to talk about the groundbreaking result at an international workshop in Finland, but the organizers wouldn't allocate him any time. He got a hearing only after he showed his calculations to Nicola Cabibbo, the Italian physicist who had introduced quark mixing in the first place. Cabibbo immediately understood the result and lobbied for Alexei to get a slot for his talk. The discovery became known as the MSW effect (for Mikheev, Smirnov, and Wolfenstein). It predicts a flavor conversion of solar neutrinos that is independent of the distance between the sun and earth. But of course this still didn't prove that neutrinos really were changing their flavor.

In the early 1990s, the solar neutrino experiments that were sensitive to the direct neutrinos from proton fusion delivered their first results. These were the GALLEX experiment led by Till Kirsten at MPIK in Heidelberg and the Soviet-American SAGE experiment located close to Mount Elbrus, the highest peak of the Caucasus. (This is a region where it is not uncommon to see dead horses rotting at the sides of the roads, where public transport buses stop until the passengers have collected money for gas, and where a village with the name Neutrino exists, with its own anthem.) Both GALLEX and SAGE used gallium as a neutrino detector, either in the form of an aqueous gallium chloride solution or in metallic form, and both of them employed quantities of it that only three decades before would have exceeded the world's production by a factor of ten. This was possible only because of the important role gallium played in the developing semiconductor industry. The price of gallium decreased greatly, but it nevertheless remained attractive for thieves, who broke into the Russian laboratory twice by using the railborne carriage leading into the underground lab (the rumbling was actually detected by the other experiments), and who stole more than two tons of the material.

The results, however, were anything but clear: While the count recorded by SAGE lay far below the expectation of solar neutrino fluxes and thus seemed to support the neutrino oscillation hypothesis, the first GALLEX result appeared to observe all neutrinos from the proton fusion and, in addition, some neutrinos produced in the subsequent solar reactions. Later the SAGE result confirmed the GALLEX result so that no clear picture emerged. Whether the earlier neutrino deficit was due to neutrino oscillations or whether it was a consequence of the limited understanding of the working of the sun could not be decided. In GALLEX, for example, the chemistry of the experiments was extremely complicated: In several steps the germanium resulting from neutrino capture had to be precipitated and extracted from the huge tanks, until the subsequent radioactive decays could be measured in tiny counting tubes and thereby indicate the capture of solar neutrinos. Finally, the MSW effect predicted a flavor conversion of neutrinos that depends on the neutrino energy. Thus even a complete observation of proton fusion neutrinos would not invalidate the MSW effect as an explanation for the disappearance of higher-energy neutrinos. In this blurry situation, experiments looking for neutrinos from a different astronomical source eventually played the decisive role.

Apart from sunlight and solar neutrinos, the earth is being bombarded also with other projectiles from outer space. To this day, the origins of the cosmic radiation are not completely understood. Most physicists believe that the most likely candidates are supernovae, the gigantic explosions accompanying the collapse of massive stars, and active galactic nuclei, black holes millions of times more massive than the sun located at the centers of galaxies, whose gravitational attraction converts their direct neighborhood into a boiling inferno of extremely energetic particles. When the quanta of the cosmic radiation (mostly protons) hit the upper atmosphere and collide with nuclei of nitrogen and oxygen atoms, medium-weight mesons—such as pions and kaons, which are bound states of quarks and antiquarks—are produced. These mesons decay quickly, creating electron and muon neutrinos in the ratio of 1:2.

Such atmospheric neutrinos were first discovered in underground experiments in South Africa and India in 1965, which provided additional hints that the observed fluxes did not correspond to the theoretical predictions. As for solar neutrinos, the situation remained unclear for a long time.

In the early 1990s several experiments, originally constructed to search for the proton decay predicted in GUT theories, delivered their most important results by measuring atmospheric neutrinos. Two of these experiments, which detected neutrinos by measuring the heat produced by them within the detector material, could find no hint of neutrino oscillations. Two others (IMB and Kamiokande), looking for the blue flashes that arise in water tanks when neutrinos hit charged particles and accelerate them to speeds greater than the speed of light in water, reported a flux of muon neutrinos about one-third less than was expected. Somewhat earlier, in 1988, just after Kamiokande had confirmed the neutrino deficit reported by IMB, John Learned of the University of Hawai'i, in collaboration with my later bosses Sandip Pakvasa (also from Hawai'i) and Tom Weiler (Vanderbilt University, Nashvillle, Tennessee), had found that the result could be interpreted as neutrino oscillations between muon and tau neutrinos with almost maximal mixing.

The final clarification came only with a rash of new experiments, and then everything happened almost at once. A third experiment that used the heat-detection method confirmed the atmospheric neutrino deficit. Then, in 1996, the gigantic

Super-Kamiokande started to take data. Super-Kamiokande—whose name, an English-Japanese amalgam, can be pronounced in Japanese to mean *super-bite-into-god*—was ten times as big as Kamiokande, and after only one year, Masato Takita reported at the Beyond the Desert conference organized by Klapdor-Kleingrothaus and me that Super-Kamiokande also seemed to confirm the atmospheric neutrino deficit. After another year the Kamiokande researchers felt confident enough to announce the discovery of neutrino mass: "This is not a small effect, and we have found no way to make it go away or even be severely distorted," explained John Learned, member of the Super-Kamiokande Collaboration and, along with Fred Reines, a founder of the IMB experiment.[9] In addition to neutrino fluxes, Super-Kamiokande was able to reconstruct the directions from whence the neutrinos came. One could actually see the position of the sun in the neutrino "snapshot" of the heavens. And this measurement provided yet another hint for the oscillation hypothesis: Only those neutrinos oscillated that had been produced in the atmosphere on the other side of the world and thus had to penetrate the entire earth on their way to the Japanese lab. The neutrinos produced above Japan had a travel distance too short for complete oscillation.

Next, experiments were built that sound like science fiction (Fig. 9.2). Neutrino beams were produced in the large accelerator labs KEK (Japan), Fermilab (United States), and CERN (Switzerland) and directed over distances of 250 km, 730 km, and 732 km, respectively, through the earth onto the underground labs Super-Kamiokande, MINOS (Minnesota), and Gran Sasso (Italy). Again the Japanese were ahead in this

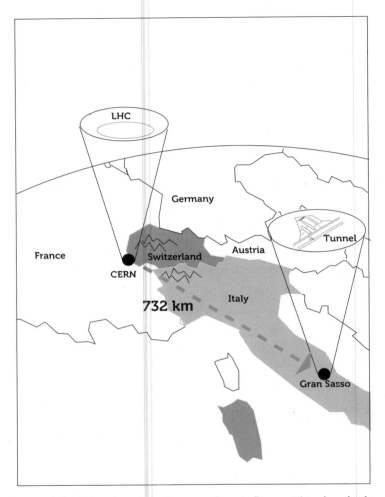

Figure 9.2. Modern long-baseline experiments fire neutrinos hundreds of kilometers through the earth aiming at distant underground labs (even one in a different country). Shown here is the beam line of the CERN–Gran Sasso experiment starting in Geneva, Switzerland, and arriving in the Abruzzo mountains, Italy, east of Rome. Here researchers could see the oscillations of muon into tau neutrinos for the first time with an earthbound source.

game by a nose. In 2006, they could observe the disappearance of muon neutrinos due to their oscillations into tau neutrinos in an earthbound experiment. And in June 2010, the OPERA experiment also detected the first tau neutrino in a muon-neutrino beam and in this way finally confirmed the atmospheric neutrino oscillations.

In 2002 the conversion of electron neutrinos into muon and tau neutrinos through the MSW effect in the sun was confirmed. It had taken fifteen years to finally complete the SNO (Sudbury Neutrino Observatory) experiment, located in Sudbury, Ontario, Canada, the delay being caused at least partly by its enormous technical complexity. To minimize the background, the physicists had to build an experiment under clean-room conditions, with the whole thing inside the caverns of a nickel mine still running at peak activity, drawing out 5,000 tons of ore every day. But when the experiment, with a thousand tons of heavy water—worth 300 million Canadian dollars and borrowed from a state-owned Canadian nuclear power company—was finally running, it nailed down the solution of the solar neutrino problem. After Kamiokande and then Super-Kamiokande had confirmed the "missing" solar neutrinos, SNO could observe not only the deficit of missing electron neutrinos but also the excess of the oscillation products, the newly born muon and tau neutrinos. This is because SNO was sensitive not only to the W-boson exchange with electron neutrinos but also to the exchange of Z bosons that is relevant for the interactions of all neutrino flavors. The solar neutrino problem was finally solved, and in the same year this result was confirmed in a terrestrial experiment when the Japanese KamLAND Collaboration measured the neutrino fluxes that emanated from several reactors about 150 kilometers away.

Again, Tom Weiler and Sandip Pakvasa, this time in collaboration with Tom's PhD advisor Vernon Barger (a PhD great-great-grandson of Enrico Fermi), were among the first to realize that the results of solar and atmospheric neutrinos require two large mixings—Tom christened this pattern *bimaximal*—and that neutrino mixing is therefore completely different from the better-known mixing of quarks, which mix only weakly.

This time, after the experiments of the 1990s had demonstrated that neutrinos really possess masses, the particle physicists immediately faced the next set of puzzling questions: Why is the neutrino so light? Why is its mixing so totally different from that of quarks? How can the neutrino be described successfully in a single theory together with quarks and charged leptons if it is at least a million times lighter? In the words of Art McDonald of Queens University in Canada, spokesman of the SNO Collaboration that finally proved neutrino flavor conversion: "What we need is a mechanism explaining neutrino masses and in particular why neutrino masses are so much smaller than the masses of the other particles."[10] Whatever these mechanisms may be, they make neutrinos ideal probes for the new physics beyond the Standard Model: for GUT theories, extra dimensions, and maybe even time travel. But before we go there, we first make a small detour into the depths of the universe.

10

Cosmic Connections

Maybe it was pigeon droppings? What was it that created this infuriating "noise" seemingly coming from all directions? Radio noise that seriously jeopardized the measurement of the galactic radio flux by the two radio astronomers Arno Penzias and Robert Wilson.[1]

An enamored pair of pigeons had chosen to nest, of all possible places, in the reflector of the radio antenna that Penzias and Wilson had gotten hold of from the Bell Telephone Laboratories. But even after the birds had been driven away— a tremendously difficult task—and the reflector had been cleaned, the problem remained. They detected radio noise corresponding to the thermal radiation of a black body having a temperature of about minus 270 degrees Celsius, about three degrees above absolute zero. Slowly Penzias and Wilson started to realize that their result had nothing to do with the galaxy, nor with the pigeons. And when the Princeton cosmology group of Robert Dicke learned, by pure accident,

about the measurement, they realized the true origin of the
noise: Penzias and Wilson had measured the cosmic microwave background, the reverberant "sound" of the Big Bang,
the cosmic wall that blocks a view into the hot primordial
phase of the early universe.

Our universe is a weird place (Fig. 10.1). Whoever gazes at
the night sky on a moonless night may see certain bright
spots that are some of the planets orbiting the sun just like
earth: Mercury, Venus, Mars, Jupiter, and Saturn (in addition
there are Uranus and Neptune, which cannot be spotted with
the naked eye). On a clear night a broad band arches across the
firmament: It is the aptly named Milky Way, our galaxy—a
spirally shaped aggregation of billions of stars, including our
sun, so dense in the night sky that when we look within the
milky "wheel" we cannot resolve the stars individually. Outside that band, but still within the Milky Way, there are many
single stars that can be spotted with the naked eye, as well as
some other galaxies, such as the Andromeda Nebula, that can
be seen as blurry luminous spots in the dark sky. The Andromeda Nebula, the small and large Magellanic Clouds, and
the Milky Way are all part of the *local group,* a galaxy cluster
about five to seven million light-years across (a light-year
being the distance light travels in a year), which in turn is
part of the Virgo Supercluster with a diameter of 200 million
light-years. Such superclusters are lined up on the largest
known scales in wiry filaments such as the "Great Wall" around
the Coma Supercluster. In between are voids billions of light-
years in diameter that contain almost no matter.

If one gazes that far into space, now with the help of powerful telescopes, one also looks into the past: The farther away

Figure 10.1. View into the depths of the universe. This "Ultra Deep Field" view by the Hubble Telescope shows an apparently almost empty region of the night sky south of the constellation Orion, an image recorded with the longest exposure time ever used. This image reveals ancient light sources: some of the very earliest galaxies, which started to re-illuminate the universe after its "Dark Ages," when the universe had cooled down but before stars had formed. (Courtesy NASA/Defense Video & Imagery Distribution System)

the observed object is, the longer the light has taken to reach us, and the older is the state of the universe displayed. At a distance of fourteen billion light-years one looks into an impenetrable wall—a wall made not of galaxies but of light. It is the wall whose leftover radiation produced the noise in Penzias and Wilson's radio antenna. Behind this wall the universe is not transparent any more. There it is an opaque plasma made of hot elementary particles. It is hiding the birth of the universe, a universe about to expand. And the wall itself is a consequence of its expansion.

When Einstein tried in 1917 to describe the condition of the universe with his theory of general relativity (Chapter 13), he couldn't find a "static" solution describing a universe changeless in time. His calculated universe either exploded or else collapsed under its own gravitational attraction. In order to achieve a static solution anyway, Einstein introduced a form of vacuum energy—the so-called *cosmological constant*—in his equations. This was supposed to avoid the collapse (or expansion) of the universe. But that calculated universe was not stable against tiny distortions. Moreover, only a decade later Georges Lemaitre and Edwin Hubble discovered that the spectra of far away galaxies showed a *red shift*: The wavelength of their light was stretched toward longer wavelengths (for visible light, that means toward the red end of the spectrum), a consequence of the cosmic expansion. The idea of vacuum energy was thus abolished, until it suddenly reappeared in 1998.

The discovery of the universe's expansion meant that one could now trace its history back to its beginnings, when it was much denser and much hotter. George Gamow, a Russian jack-of-all-trades who followed his interests from nuclear physics to cosmology and later to molecular biology, joined with Ralph Alpher and Robert Herman in the 1940s to predict the existence of the wall—the cosmic microwave background. It emerges when, after about 300,000 years of expansion, the universe has cooled down enough that electrons and atomic nuclei can bind together into atoms. Suddenly, the universe becomes electrically neutral and optically transparent. The remaining light quanta can spread undisturbed, to get detected today—red-shifted into microwaves from the many billion years of cosmic expansion. The discovery of Penzias and Wilson was spectacular, as it proved the birth of the universe in a hot fireball: the Big Bang. But what can be found on the other side of the wall? What do we know about the true origin of the universe?

When, in 2005, the website boston.com announced a competition for the room most urgently in need of a spring cleaning, Alan Guth's office at MIT was the unrivaled winner: "Now, we've seen unorganized, cluttered, and messy, but the office of Alan Guth takes the cake in all categories."[2] It's ironic that of all people the winner was Alan Guth, the man who had cleaned up the whole universe. In 1978—twenty-seven years earlier—Guth was a young postdoc at Columbia University. He had published only a modest number of papers on particle theory and didn't nurture an intense interest in cosmology. In the fall of that year, however, he attended a seminar by Robert Dicke, the man who, in 1964, had helped Wilson and Penzias understand their discovery. Dicke talked about a besetting problem of Big Bang cosmology: The universe is surprisingly flat! According to general relativity, space-time is, for example, warped around the sun. Overall however, the universe doesn't exhibit much curvature. Quite the contrary, the universe is *extremely* flat; deviations of total flatness amount to less than a few percent today, corresponding to less than one part in 10^{60} at the Planck scale in the early universe.

Guth was fascinated, and, serendipitously, on the very same day over lunch, Henry Tye, who had earned his PhD together with Guth at MIT, drew Guth's attention to another problem: The universe is surprisingly empty.

If the universe was so hot at the time of its birth that temperatures beyond the GUT scale were reached, exotic objects such as magnetic monopoles should have arisen in large numbers during the spontaneous breaking of a GUT symmetry at the boundaries of different vacua, meaning different ground states of the Higgs field forming in different regions of the universe. All known magnets have two poles, a north pole and a south pole. Even if broken, every piece separately acquires both poles. An object with a single north or south pole has never been found, although gigantic experiments such as MACRO in the Gran Sasso lab look for them. Well, strictly speaking, exactly one candidate for a monopole signal had been reported—on Valentine's Day 1982 by Blas Cabrera in California. If Guth's scenario of cosmic inflation, of which we are about to learn, is correct, this might have been the only monopole in the entire observable universe. Or it could have been an error in measurement.

In any case, Tye convinced Guth that they should work together on the monopole problem, and their first paper was almost finished when Guth, sitting late at night at his desk, checked some of Tye's calculations and made a spectacular discovery: The energy due to vacuum fluctuations drives the universe to expand faster and faster and eventually even beyond light speed—*exponentially* fast. The scenario that was worked out in the following years by Guth, Andrei Linde, Paul Steinhardt, and others, is called *inflation*. The field responsible for the vacuum energy that triggers the process was christened the *inflaton*. By means of an inflationary phase in the early universe, the monopoles become diluted. In an instant the universe is deserted and tidy, in stark contrast to Guth's office.

At the same time, inflation provides a solution for the flatness problem: Just as the surface of the earth appears flat in a local region even though the earth is a sphere, the universe gets so radically blown up that that our accessible neighborhood appears flat.

The idea that hit him that night made Guth famous and, if not rich, at least secure: Not more than two months after his nocturnal discovery, Guth gave a talk at Rutgers University in New Jersey, where the inventor of quarks, Murray Gell-Mann, jumped up and shouted: "You have just solved the most important problem in the entire field of cosmology."[3]

Only a little later, Guth received job offers from some of the best universities in the world and chose MIT. But the physics would get even better! As it turned out, inflation was capable of solving two more of the fundamental problems of cosmology (Fig. 10.2).

- The matter in the universe is surprisingly smoothly distributed.

If one considers the universe on sufficiently large scales, it looks everywhere pretty much the same. In particular the cosmic microwave background radiation shows only tiny fluctuations around its mean temperature (Fig. 10.3). This uniformity applies even to regions that, according to the older version of Big Bang cosmology, should never have been in causal

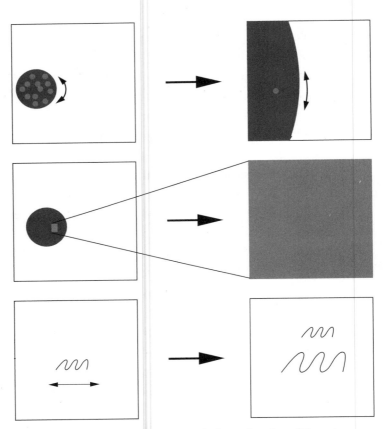

Figure 10.2. A primordial phase of inflation solves four different problems of cosmology. Upper picture: When space gets blown up the curvature is diminished (space becomes "flatter") and exotic relics such as magnetic monopoles (grey dots) are diluted. Center picture: A small region in space that was small enough to establish a balanced temperature at an earlier time in the history of the universe is being expanded so strongly that it contains the entire observable universe. Lower picture: Quantum fluctuations cannot revert back within this rapidly expanding space-time (the so-called freeze out) and therefore get stretched. New quantum fluctuations during the inflationary phase then look like small copies of the earlier fluctuations. The pattern of inhomogeneities is identical on different magnitude scales.

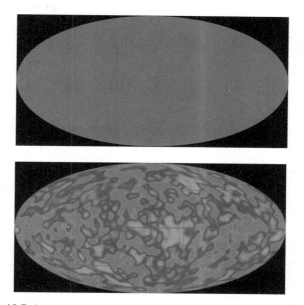

Figure 10.3. A cosmic-microwave-background snapshot of the early universe. At first glance all of the microwave radiation seems to have the same temperature (upper picture). Only when temperature fluctuations 100,000 times smaller than the average temperature are resolved (lower picture) do inhomogeneities appear. This structure in the radiation is assumed to arise from density perturbations in the early universe created by quantum fluctuations during the inflationary phase. (Courtesy Lawrence Berkeley National Laboratory)

contact with each other. (This is also known as the *horizon problem*.) The superluminal expansion during inflation can then explain the fact that regions in the universe that were originally in causal contact appear causally separated today.

- The uniformity isn't perfect. The universe contains structures such as galaxies, galaxy clusters, superclusters, and filaments.

As soon as one starts to look at these deviations from uniformity, one at once makes another surprising observation: The universe looks very much the same at different magnitude scales. The deviations from uniformity resemble each other, whether one looks at the density fluctuations of galaxy clusters or superclusters. This means that one finds tiny fluctuations over small distances (small, that is, by cosmological standards), small fluctuations over larger distances, and larger fluctuations over huge distances. This pattern of density fluctuation arises automatically during a phase of inflation (Fig. 10.2), since vacuum fluctuations cannot revert back due to the superluminal expansion and thus are carried away and stretched as lumps of energy. Since this happens constantly, while the space below the fluctuations expands, a pattern of inhomogeneities forms that looks the same on all magnitude scales.

In this way inflation provides the birth of the universe, and at the same time the origin of all structure. At some point the inflaton field decays into the known particles. Wherever the fluctuations of the field produced a larger than average energy density, there was stronger gravity, whose attraction caused more and more particles to accumulate. Later on such accumulations grew into galaxies, clusters, superclusters, and filaments. As Rocky (Edward W.) Kolb, a cosmologist at the University of Chicago, puts it: "Everything originates from inflation: from the biggest galaxies to the lowest life forms." With the last words he likes to flash pictures of famous politicians.

Despite all that it has going for it, inflation is so far only a scenario. There exists no totally satisfying particle-physics model for the inflaton field. "Who is the inflaton?" Kolb pointedly asked at the 1997 Beyond the Desert conference.

Moreover, it is extremely difficult to make meaningful statements about the condition of the universe before inflation, since inflation erases all traces of any such preexisting state. It is not even known whether a Big Bang happened at all before inflation. Possible, and perhaps even more probable, is the concept of eternal inflation, in which new baby universes keep popping up like bubbles out of nothing. Only in a few of these universes does the inflaton field decay into a hot plasma of particles, wherein inflation ends and the post-inflationary plasma looks just like the hot soup of particles emanating from a Big Bang.

It's not just the *origin* of the universe that is still not completely understood. The universe as it exists today is an extremely puzzling thing. Only 0.5 percent of the matter and energy content of the universe radiates light that can actually be seen, and only 5 percent consists of matter that we understand (mostly hydrogen and helium). Furthermore, we don't know where that 5 percent came from.

What Is Dark Matter?

Astronomical measurements such as the rotation rates of stars in spiral galaxies show that the gravitational pull of the mass of luminous celestial bodies in the galaxies is, by itself, not sufficient to keep the stars in their orbits. Additional dark (nonluminous) matter (or energy) is necessary and contributes to the total mass of the galaxy. This result is supported by additional cosmological observations. About 23 percent of the energy of the universe consists of this puzzling dark matter. At the same time, it is excluded that this mass is made out of bound states of known elementary particles. It is rather assumed that heretofore unknown particles constitute the dark matter. Such particles should be electrically neutral and thus have properties similar to those of heavy neutrinos. Indeed new neutrino species are being discussed as possible candidates for dark matter. The most popular dark-matter candidate at present is the lightest supersymmetric (SUSY) particle, if it is prohibited from decaying into lighter Standard Model particles and thus remains stable.

What Is the Dark Energy?

Even more puzzling than dark matter is the so-called *dark energy*. After Lemaitre and Hubble had concluded in the 1920s that the universe is expanding, Einstein removed the cosmological constant from his equations and reportedly called it his "biggest blunder."

It would take some seventy years, until 1998, before a measurement of the red shift of far-off supernova explosions demonstrated that the present expansion of the universe is accelerating—a fact that doesn't fit the model of a universe being decelerated by the gravitational pull of its own matter. Einstein's cosmological constant, a kind of vacuum energy—now denoted as dark energy—was reborn and now contributes the major part (about 73 percent) of the mass of the universe.

The origin of this vacuum energy is so far totally unknown. In particular it is puzzling why the vacuum energy is so small. If one calculates typical contributions to vacuum energy, coming, for example, from Higgs fields (there might be more than one) and quantum fluctuations, the result is about 120 orders of magnitude larger than the present value. That is a one followed by 120 zeroes. As successful as the Standard Model is in other respects, this is probably the worst prediction in the history of science. For a long time, until supernova data provided evidence of accelerated expansion, physicists believed that an unknown mechanism would cancel the vacuum energy, making it exactly zero. Now we have hints of a new era of inflation. Physicists are still racking their brains trying to find an explanation for why the vacuum energy is non-zero but still so tiny.

How Was Matter Created?

The German existentialist Martin Heidegger, infamous even more for his incomprehensible metaphors than for his enmeshment with the political system of Nazi Germany, declares in his magnum opus *Being and Time:* "nothingness annihilates itself."[4] Above all, nothingness annihilates itself in cosmology, as it is far from simple to explain why there is anything at all in the universe rather than nothing. In the first place, there is no good reason why the decay of the inflaton should produce matter but not the same amount of antimatter. In that case matter and antimatter would annihilate, and the consequence would be a universe consisting only of radiation, without matter or structure. And even if an excess of matter existed at the beginning of time, this excess would have been diluted so strongly in the process of inflation that it would have been almost completely erased. In experiments that were suspended from a balloon flown forty kilometers up into the atmosphere above Antarctica, and with spectrometers aboard the spaceship Discovery as well, experimenters searched for antiprotons and antinuclei in the cosmic radiation—without any success. Moreover, telescopes do not find any sign of the annihilation radiation that would be expected to be generated at the boundaries between large regions filled with matter and antimatter. This means that there has to exist a mechanism that transforms antimatter into matter, or which generates more matter than antimatter in the first place. Or more specifically: more baryons than antibaryons. Such a mechanism is called *baryogenesis*. For the universe to produce more baryons than antibaryons, the underlying process has to fulfill three conditions first formulated in 1967 by the Soviet nuclear physicist, cosmologist, dissident, and Nobel Peace Prize laureate, Andrei Dmitrievich Sacharov:

- The conservation of baryon number has to be violated. Otherwise baryons and antibaryons could not be produced in different numbers or transformed into each other.
- Particles need to behave differently from antiparticles. If the process works for particles and antiparticles in the same way, nothing is gained in the end. That means that the C and CP symmetries—charge conjugation (exchange of particles and antiparticles) and charge conjuga-

tion in combination with mirroring—have to be violated.

• The process must run in one direction only, not back and forth with the same strength. Physicists call this *nonequilibrium*.

But what has all of this to do with neutrinos? A lot. The tiny neutrino, despite its minuscule mass, is one of the global players in cosmology. Just as the cosmic background radiation is a relic of the photons in the early universe, there also has to be a relic neutrino background. Three hundred thirty neutrinos per cubic centimeter, spread throughout starless realms as well as galaxies, should now fill the universe as relics of its hot birth. While the relic neutrino background hasn't been observed yet, the most promising proposal for such a detection has been advanced by Vanderbilt University's Tom Weiler: High-energy neutrinos created in astrophysical explosions, in collisions with the relic neutrino background, could produce Z bosons, which would then decay into showers of secondary particles, observable in experiments looking for cosmic radiation. An advantage of this *Z-burst* process, as it is called, is that it depends on the neutrino mass, so that as a by-product it reveals additional information that can be compared with the results of neutrinoless double beta and tritium beta decay. (Tom Weiler and I performed such an analysis in 2001.) The drawback is that there are not enough high-energy neutrinos arriving in our cosmic neighborhood to make it possible to observe Z-bursts in present-day experiments.

Yet neutrinos are so ubiquitous and so numerous in the universe that if the sum of their masses exceeded 40 eV (which is less than one ten-thousandth of the mass of an electron) they would have made the universe re-collapse long before it reached its present size and before humans could have evolved. So we owe our existence to the lightness of neutrinos! But neutrinos potentially have influenced the fate of the universe and may have been crucial for our very existence in many other ways. For instance:

• *Big Bang nucleosynthesis:* The creation of light atomic nuclei.

We are all made out of stardust. Really. Indeed, elements such as carbon, silicon, oxygen, and nitrogen, which build the structures of flesh, blood, and bones, of leaves, rock, and air, have been synthesized from even lighter elements inside early stars and later blown into space when those stars ended their lives in supernova explosions.

The lightest elements, on the other hand, notably hydrogen and helium, were just as romantic, having been created within the first three minutes of our universe. The correct prediction of their relative abundance is one of the most convincing pieces of evidence that our universe was born in a hot fireball. This relative abundance depends strongly, in fact, on the number of neutrino flavors. While the Standard Model contains only three neutrino flavors—one per family—additional flavors can arise in extensions of the Standard Model. More neutrino species means more fast particles (or "radiation") in the early universe, which forces the universe to expand faster.

This in turn means that protons and neutrons have a shorter time in the early universe to transmute into each other in collision processes with other particles. Beyond a certain volume of space the probability to hit another particle of the correct type is simply too small, and the ratio of protons to neutrons remains constant beyond that volume, apart from the slowly proceeding decays of neutrons. When that happens earlier, meaning at higher temperatures, or equivalently higher particle energies, the small difference in mass between neutron and proton, which tends to shift the ratio toward a higher concentration of the lighter protons, has a negligible effect. More neutrino flavors thus imply more neutrons and thus more helium, which consists of protons and neutrons, whereas hydrogen (or at least its most abundant isotope) has a single proton in the nucleus. The correctly predicted relative abundance of hydrogen and helium in the present universe is based on there being just three neutrino flavors. If there are more than three, the additional ones were evidently not produced in the early universe.

- *Structure formation:* How did the original inhomogeneities in the inflaton field evolve into structures such as stars, galaxies, clusters, and filaments?

When the universe cooled down, heavy, cold dark matter particles got pulled into the slowly accumulating matter lumps. At this time neutrinos were still hot and raced almost freely across space. Although these streaking neutrinos did not themselves cluster, they could tear apart lumps of cold dark

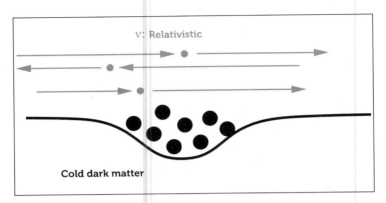

Figure 10.4. Erasure of structures in the early universe by neutrinos. While the cold, heavy dark-matter particles lump together, the light, hot neutrinos tear the lumps apart.

matter that already had formed via their gravitational pull. The larger the neutrino mass, the more effectively structures were erased on length scales corresponding to the distance neutrinos could stream freely (Fig. 10.4).

Astronomical observations of the structures in the universe and measurements of the cosmic microwave background radiation in which the structures are imprinted therefore allow conclusions about neutrino masses. Analysis of the data leads to the conclusion that the neutrino masses are less than a few tenths of an eV (meaning at most one millionth of the mass of an electron). Otherwise the structures could not have formed as observed. This value is smaller than the present sensitivity of double beta decay and tritium beta decay experiments. These terrestrial experiments are nevertheless worth pursuing, because there are many complicating factors in the cosmological measurements and analysis that make the calculated neutrino mass limit rather uncertain.

Dark Matter

As already mentioned, it is possible that new, heavy neutrinos can themselves be the dark matter. But even if dark matter is made up not of neutrinos but of SUSY particles, these would accumulate, because of their gravitational attraction, in the core of the earth or the sun and there annihilate into neutrinos. This high-energy neutrino radiation would spill out from the interior of the earth or the sun and could be measured by the huge neutrino telescopes that, as of this writing, are under construction in Antarctica and the Mediterranean (see Chapter 16). This process is known as *indirect detection,* and, along with *direct detection* of dark matter particles flying through a detector in an underground lab (Chapter 8) and *direct production* of dark matter particles in an accelerator, is a promising strategy in the search for the biggest matter component in the universe.

Dark Energy

Vacuum energy and neutrino masses have at least one thing in common—their puzzlingly small magnitudes. Whether there really exists a relation between dark energy and neutrinos is so far unknown—as is the origin of dark energy itself.

There are, however, some interesting facts that deserve notice. One is that the energy density of dark energy is in the same ballpark as the neutrino mass to the fourth power, when *natural units* are used, meaning the unit of volume is understood to be inverse mass to the third power. P. Q. Hung of the

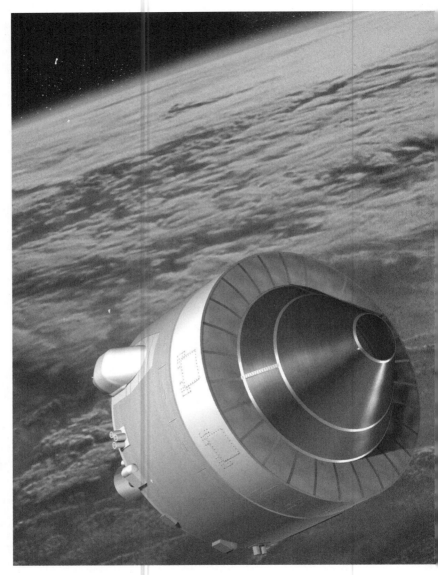

Figure 10.5. The PLANCK satellite of the European Space Agency (ESA) surveying the cosmic microwave background radiation with an as-yet-unknown accuracy. It may reveal new information about neutrino masses and the mechanism behind inflation. (Courtesy ESA—D. Ducros)

University of Virginia noticed this, and he has proposed that the vacuum energy of the field that provides masses to neutrinos is also responsible for the dark energy. A similar model had also been proposed previously by Anupam Singh, at that time at Carnegie Mellon University. In 2003 Ann Nelson, Robert Fardon, and Neal Weiner published a more elaborate model in which dark energy is regulated by the neutrino mass density, and in which the ratio of the contributions of neutrinos and dark energy to the total energy of the universe remains constant over a long era in the evolution of the universe.

Such models predict that the neutrino mass is not constant, but changes with the evolution of the universe so that, as P. Q. Hung and I showed in 2003, MSW-like transitions among neutrino flavors can also arise in empty space.

Another interesting link between neutrinos and dark energy is *the dark-energy seesaw.* As we shall see in the next chapter, the smallness of neutrino masses can be explained with a so-called *seesaw mechanism,* which suppresses the actually expected mass (of the order of the typical masses of charged leptons and quarks) by the large mass of a GUT-scale particle. Similarly, a lower bound for the expected vacuum energy is the energy scale where SUSY particles can be produced for the first time. (At higher energies quantum fluctuations of particles and SUSY partners can cancel each other, just as I discussed in Chapter 6 for the quantum fluctuations contributing to the Higgs mass.) Although there doesn't exist an accepted explanation for this observation, it is intriguing that, similar to the relation of expected value and actual observed limits of the neutrino's mass, the observed value of the dark energy density just corresponds to this SUSY scale squared, divided by the Planck energy scale, which is the large energy scale of quantum gravity and where dark energy might originate.

Accidental, or not? If nothing else, at least interesting.

Baryogenesis and Inflation

Neutrinos can play a crucial role in baryogenesis, the generation of heavy matter in the universe, and may also be involved in driving inflation. To understand these ideas we will deal with models for neutrino mass generation in the next chapter.

In summary: Neutrinos influence cosmology and cosmology influences our understanding of neutrinos. Indeed cosmology became a precision science in recent years, and is now in a phase as exciting as particle physics.

The next exciting result will probably be a more accurate determination of the cosmic microwave background radiation by the ESA satellite PLANCK, launched in May 2009 (Fig. 10.5). PLANCK aims to improve the cosmological sensitivity to neutrino masses to values below 0.1 eV. Finally, the PLANCK measurement of the tiny inhomogeneities in the cosmic microwave background radiation will help to constrain and improve understanding of the mechanism of inflation.

11

Neutrinos

Key to the Universe

The restrooms of the downtown Dortmund pub Kraftstoff (meaning "fuel") are wallpapered with a superhero cartoon, in which the protagonist announces stone-faced: "I'm looking for a very rare key, made out of the finest crystal and hidden in a deep vault." This picture is an ideal metaphor for neutrino physics in the new millennium. Neutrinos are indeed sought in the deep vaults of underground labs, and they indeed could serve as keys for GUT theories—or even more exotic theories—to explain the physics of elementary particles. The link between neutrino masses and the deepest mysteries of the particle world is the mechanism of neutrino mass generation: the answer to why the neutrino mass is so special. The discovery of neutrino mass in the 1990s provided the theorists with two rows to hoe. First: How can neutrino masses be implemented in the Standard Model? Second: Why are neutrinos so much lighter than the other elementary particles such as quarks or the electron and its heavier siblings, muon and tau? And how can neutrinos be described successfully in a unified theory together with the charged leptons and quarks if they are at least a million times lighter?

To clarify these questions, particle theorists have proposed several models of neutrino mass generation. The most popular explanation is known as the *seesaw mechanism* and uses a logic that appears rather strange at first glance: The observable, left-handed neutrino is so light because its right-handed partner is so heavy. The idea behind this is that the left-handed neutrino, in order to generate a mass, has to fluctuate temporarily into the heavy right-handed state. Since the probability of this quantum fluctuation, which relies on the energy version of the uncertainty principle (see Chapter 4), is small when the energy or mass that has to be "borrowed" from the vacuum is large, the process is extremely unlikely and the resulting mass is small. As a metaphor one could imagine a mountain being so high that most mountaineers are discouraged from trying to climb it and an ascent is reported only rarely.

Why is this explanation so popular among physicists? One reason is that it allows one to connect the neutrino with fundamental puzzles of particle physics such as the realization of a unified GUT theory or the question of why the universe contains almost exclusively matter, and not matter and antimatter in equal parts. Beginning in the year 2000, a downright industry of theoretical research groups sprang up whose members studied the interrelations among GUT physics, neutrino masses, and observables such as lepton-flavor-violating decay rates. Toilers in this industry included J. Alberto Casas and Alejandro Ibarra in Madrid, Sergey Petcov and Werner Rodejohann (at that time in Trieste), and Frank Deppisch,

Andreas Redelbach, Reinhold Rückl, and me (at that time all in Würzburg). The other reason is that the seesaw mechanism arises quite naturally as soon as one tries to endow the Standard Model with neutrino masses. Recall that the neutrino is special among the elementary particles in two important respects:

- Neutrinos are the only electrically neutral particles among the elementary fermions, or matter particles.
- Neutrinos are far lighter than the other matter particles.

The seesaw mechanism (as well as most other mechanisms proposed for neutrino mass generation) postulates a relation between these two special properties: It assumes that the neutrino is so light because it is neutral.

Let's remind ourselves of Majorana's last work, which he wrote shortly before he disappeared from the ferry to Palermo: Since the neutrino doesn't carry any charge, it is possible that it is its own antiparticle. But things are not that easy. The left-handed neutrino that we know from the Standard Model and that can be produced and annihilated in weak-interaction processes doesn't carry any *electrical* charge, but it carries the generalized charge of the weak interactions, isospin. We remember how the masses of the electrically charged leptons and quarks are generated: A mass corresponds to a transition from a left-handed into a right-handed state (or the other way around). For quarks or charged leptons, this transition is possible only because the particles are coupled to the

Higgs, which absorbs or delivers the difference of the weak charges of the left-handed particle and its weakly uncharged right-handed partner. This is the reason why masses can't become much larger than the vacuum energy of the Higgs field. If one wanted to generate a mass that transforms left-handed neutrinos into right-handed antineutrinos, including the two Standard Model states that are observable in experiments, both of these states would have to shed isospin charge. To make this possible one would need a novel Higgs field that could absorb the isospin excess, a so-called *Higgs triplet.* This possibility is now known as *seesaw-II.*

There is in fact a simpler possibility for generating neutrino masses—adding a right-handed neutrino to the Standard Model particle roster. In order to endow neutrinos with masses, they must be able to transmute into right-handed particles. It therefore seems natural to assume that in analogy to the mass generation of charged leptons and quarks, a right-handed neutrino exists, and that the left-handed neutrino can be transformed into it via interaction with the Higgs field. This right-handed particle can now really be a Majorana particle, meaning its own antiparticle, since it carries neither electric charge nor color charge nor isospin: It is totally neutral, or *sterile.* This right-handed particle can thus possess a mass that transforms the right-handed neutrino into a left-handed antineutrino. Both are equally sterile, meaning no charges obtained from the Higgs field are necessary to make this metamorphosis possible. The resulting mass can then be much larger than the vacuum energy of the Higgs field, which is not needed at all. It typically is as large as the energy where a more extensive symmetry, such as a GUT symmetry, shows up. Such a symmetry may create novel forces—such as lepto-quark forces under which the right-handed neutrino now *is* charged eventually—thus setting the cutoff for the right-handed neutrinos just as the Higgs does for the normal masses of charged leptons and quarks. Such an extremely heavy right-handed neutrino (or left-handed antineutrino, that is) can only be produced from an extremely short-lived quantum fluctuation, and it immediately decays back into a right-handed antineutrino with another coupling to the Higgs field (see Fig. 11.1).

The logic following from this chain of arguments seems rather compelling, and the consequences are breathtaking:

- The known left-handed neutrinos are so light because they acquire their masses from a short-lived quantum fluctuation into an extremely heavy right-handed neutrino.
- Neutrinos may carry information about the physics at the GUT or even at the much higher-energy string scale, which could be approached directly only with accelerators having a circumference of a trillion kilometers or more.
- A possible answer to the question Why is there something rather than nothing? or Why does the universe contain matter, and not matter and antimatter in equal amounts that would annihilate in a gigantic explosion?
- By taking a detour through heavy right-handed neutrinos, left-handed neutrinos can turn themselves into right-handed antineutrinos. This

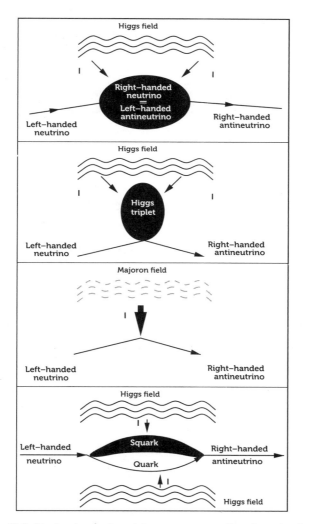

Figure 11.1. Mechanisms of neutrino mass generation, from top to bottom: (1) Simple seesaw, meaning fluctuation into a heavy right-handed neutrino. (2) Seesaw II: radiation of a quantum fluctuation of a heavy Higgs triplet. (3) Coupling with a majoron field. (4) Fluctuation into a SUSY particle, the squark, and a quark.

metamorphosis can then cause neutrinoless double beta decay.

The credit for the discovery of the seesaw mechanism is often given to Murray Gell-Mann (father of quarks), Pierre Ramond (father of superstrings), and Richard Slansky (who worked in group theory); and independently to Tsutomu Yanagida, who later also invented *leptogenesis*, a mechanism for generating the matter excess in the universe by neutrino decays. When, in 2004, a workshop celebrating the twenty-fifth anniversary of the seesaw mechanism was held at the Institut Henri Poincaré in Paris, however, almost every second person in the audience seemed to have a claim for the discovery of the mechanism, among them Sheldon Glashow (Boston University), father of the Standard Model and the first GUT theory; Rabi Mohapatra (Maryland); Goran Senjanovic (ICTP Trieste), who also introduced the seesaw-II mechanism; and Jose Valle (Valencia) and his PhD advisor Joel Schechter (Syracuse University), who in 1980 had developed a very general representation of neutrino mass generation. As it turned out, though, the very first who had speculated about the seesaw mechanism—in fact even two years earlier than the title of the workshop suggested—was the GUT developer Peter Minkowski of Bern University in Switzerland. Minkowski, however, being too relaxed to seriously fight for credit, didn't participate in the 2004 meeting in Paris, and had almost forgotten his discovery, or at least forgotten in which of his papers he had published it.[1]

In any case, the Standard Model has to be extended in order to explain neutrinos' masses—leading to the wonderful consequence that neutrino physics allows one to obtain information about GUTs or other theories of unification that are realized at energy scales inaccessible to accelerator experiments. As popular as the seesaw mechanism is, there naturally are alternatives (Fig. 11.1). The *inverse* (a.k.a *double*) *seesaw,* which was proposed in 1986 by Rabi Mohapatra and Jose Valle, utilizes a ladder of two neutrino particles without isospin into which the original neutrino oscillates, one after the other. The fact that the two sterile neutrinos can share the suppression between themselves allows, among other things, for a significantly lower mass scale of the right-handed neutrinos, perhaps at the TeV scale, which could be probed at the LHC.

Another idea was proposed in 1980 by Yuichi Chikashige, Rabi Mohapatra, and Roberto Peccei (at that time all in Munich, and now scattered to Tokyo, Maryland, and UCLA): A symmetry forbids the violation of lepton number, and only a spontaneous and small breaking of this symmetry eventually allows for small neutrino masses. The particle that plays the role of the Higgs particle breaking the lepton-number symmetry in such scenarios is called a *majoron*. It can be emitted in neutrinoless double beta decays. Moreover, in this model, neutrinos are unstable and can decay into majorons, which, by carrying away energy, crucially affect the process of supernova explosions. In my diploma thesis I analyzed double beta decay data to search for majoron-emitting decays. Later, Jose Valle, Ricard Tomas, and I compared the sensitivity of double beta decay and supernova explosions to the size of majoron-neutrino couplings, building on a previous work by Michael Kachelriess.

Finally, neutrinos could acquire their masses from quantum fluctuations into two or more particles, a so-called *particle loop*. Such models typically assume that the generation of neutrino masses via fluctuation into a single particle is forbidden, while the fluctuation into a pair of particles results in another suppression factor of about one hundred. The first model of this type was proposed by Tony Zee in 1980. Particularly interesting, however, are such models within the context of supersymmetry. In SUSY models, lepton- and baryon-number-violating interactions that violate the so-called *R-parity* (see Chapter 6) arise naturally. These interactions allow the neutrino to fluctuate into a loop including a SUSY particle and its Standard Model partner, as found out in 1984 by Lawrence Hall and Mahiko Suzuki. Moreover, in such SUSY models neutrinos can acquire their masses also by mixing with zinos, the SUSY partners of the Z bosons, as proposed in 1987 by Arcadi Santamaria and Jose Valle, building on works by Charanjit Singh Aulakh, Rabi Mohapatra, and others. This again leads to a seesaw-like formula. In these models neutrinos do not deliver information about GUT or string theories, but about supersymmetry instead. In particular, the Valencia group around Valle, Santamaria, Martin Hirsch, and Werner Porod—since the 1990s one of the most important centers of neutrino physics worldwide—has studied extensively the relations between R-parity violation and neutrino masses and has found, among other things, interesting signals at colliders such as the LHC. But the violation of R-parity has also been intensely studied by the Bonn group in Germany around Herbi Dreiner and Manuel Drees, and the Kolkata group with Gautam Bhattacharyya and Palash Pal.

This argument has also been turned around by Bhattacha-ryya (who visited our research group in Heidelberg in 1997 and 1999), Klapdor-Kleingrothaus, and me, when we used the smallness of neutrino mass to constrain the possible strength of R-parity-violating interactions. The upper bounds we found in this way improved previous constraints by a factor of a million. Above all it is Jose Valle and Rabi Mohapatra, who deserve credit for carrying out, since the 1990s, the important task of tirelessly reminding the particle-physics community about the crucial role that neutrino physics can play in the quest for new physics beyond the Standard Model. Around the end of the 1990s, these two were joined by Manfred Lind-ner; out of his group at least seven independent major re-search groups for neutrino physics have evolved.

Eventually almost all neutrino-mass models can be under-stood as different realizations of an image shaped by Steven Weinberg in 1979: A neutrino fluctuates—via a twofold inter-action with the Higgs condensate—into something heavy and then back into its own antiparticle. The neutrino becomes a Majorana particle. Particle and antiparticle are connected via their mass, and the isospin excess migrates into the Higgs field. The entire process generates a light neutrino mass, sup-pressed by the unspecified heavy object in the intermediate state.

An important reason why the simple seesaw remains the most popular mechanism of neutrino mass generation is, besides its naturalness, its cosmological relevance. Tsutomu Yanagida was one of the first to discuss the seesaw mecha-nism as an explanation for the small neutrino masses. To-gether with Masataka Fukugita, he then found in 1986 a spectacular application of the seesaw to generate the baryon asymmetry of the universe. Originally it had been assumed that the leptoquark particles that arise in GUT theories and which can cause the conversion and decay of baryons and leptons into each other were responsible for baryogenesis. As it turned out, though, the universe would have had to reheat strongly after inflation in order to reach the necessary temperatures for GUT leptoquarks to be produced. This would, however, also bring about a substantial production of weakly interacting SUSY particles, whose decays would disrupt a successful nucleosynthesis—one of the cornerstones of cosmology.

Another idea goes back to a discovery made by the Dutch Nobel Prize winner Gerard t'Hooft in 1978: Within the Standard Model there are transitions between different vacuum states that correspond to distinct Higgs condensates. These transitions can transmute antileptons into baryons and in this way create a baryon asymmetry as proposed in 1985 by Vadim Kuzmin, Valery Rubakov, and Mikhail Shaposhnikov. The energy necessary for such a transition is known as *sphaleron*. The hope was that the condensation of the Higgs field at energies of around 1 TeV would proceed so suddenly that an imbalance between baryon production and baryon annihilation would result. As it turned out, though, the mechanism doesn't work in the Standard Model, where the Higgs condensation process happens slowly, not suddenly.

Baryogenesis with the help of neutrinos, the so-called *leptogenesis* of Yanagida and Fukugita, combines the idea of particle decay with baryogenesis as a result of sphaleron transitions at low energies. The right-handed neutrinos are identical to their antiparticles. Their mass corresponds to the transmutation of particles into antiparticles. Thus the right-handed neutrinos could indeed be the reason why the universe contains almost no antimatter: In the early universe there may have been particles and antiparticles in equal number. For the right-handed neutrinos, however, there was no way to decide whether they were particles or antiparticles. When the universe cooled down, then, the temperature became insufficient to produce these heavy neutrinos and they decayed a little more often into antileptons than into leptons. Thereafter almost all particles and antiparticles annihilated, with only the tiny excess of antileptons in the decay products of the heavy neutrinos surviving. These were transmuted via sphaleron transitions into baryons, which account for almost all matter in the present universe: We were all neutrinos once.

In this scenario, neutrinos are directly responsible for the existence of matter in the universe. In later work, leptogenesis was worked out in more detail by my Dortmund colleagues Marion Flanz, Emmanuel A. Paschos, and Utpal Sarkar, and also by Wilfried Buchmüller and Michael Plümacher at DESY Hamburg and by Apostolos Pilaftsis (at that time in Munich and today in Manchester). It is now the most popular proposal for the generation of a baryon asymmetry and provides yet one more example of the relevance of neutrinos in cosmology. Moreover, for successful leptogenesis, not even lepton-number violation is necessary. As Evgeny Akhmedov, Valery Rubakov, and Alexei Smirnov, as well as Karin Dick, Manfred Lindner, Michael Ratz, and David Wright found out, a lepton asymmetry in the left-handed sector can be balanced by a corresponding lepton asymmetry in the hidden, noninteracting right-handed sector. In this case the mechanism works even for Dirac neutrinos.

What is probably the most radical proposal for the role of neutrinos was put forward in 1993 by Hitoshi Murayama, Hiroshi Suzuki, Tsutomu Yanagida (at that time all at Tohoku University, Japan), and Jun'ichi Yokoyama (University of Kyoto): The SUSY partner of the neutrino, the sneutrino, they suggested, could be the inflaton. In the decay of the inflaton field, the sneutrino could then directly decay more often into antileptons than leptons, and thus generate a lepton asymmetry, which later would be transformed into baryon asymmetry via sphaleron transitions. An extension of this idea, which allows the neutrino to be incorporated in a better way in GUT theories, has been worked out by Stefan Antusch and collaborators.

According to the ideas outlined in this chapter, the neutrino could be the reason we exist—indeed it may even be the cause of the existence of the entire universe.

Perhaps not quite as important, but similarly mind-boggling, is another speculation about how neutrinos might be related to the universe at large: the possible existence of extra dimensions. That is a topic for the next chapter.

Extra Dimensions, Strings, and Branes

The San Francisco Bay Area is a magical place. Its Golden Gate Bridge (Fig. 12.1), a symbol in steel and an object of projection for dreams, stretches across the bay, sometimes—depending on the light—a lucent red. It's an area that, over the years, has irresistibly attracted gold diggers, Beat poets, hippies, gays, computer nerds, oddballs, and creative people of all kinds—men and women who dreamed of a life outside of all social conventions where they could freely follow their ideas, impulses, and affections, and could try out whatever seemed attractive. New ideas and creativity are also the fuel of science, so it's not surprising that some of the best research institutes in the world are located around the bay. Among them there are two particularly impressive temples of science, the University of California, Berkeley, on the east bank, with its Lawrence Berkeley National Laboratory, where Hitoshi Murayama speculates on whether neutrinos could have been the origin of the universe; and on the opposite side of the bay, about thirty miles southeast of San Francisco, Stanford University, where Andrei Linde and Lenny Susskind argue about

Figure 12.1. Symbol of the California Dream: the Golden Gate Bridge.

parallel universes and string theory, and which is home to the linear collider center, SLAC.

In the fall of 1997 a young PhD student moved across the bay from Berkeley to Stanford.[1] The puzzle that kept Nima

Arkani-Hamed awake at night was the question of why gravity is so weak. Admittedly, gravity in daily life may well have fatal consequences—for example, if you fall off a roof. Still, gravity is far weaker than the other three forces: the strong, electromagnetic, and weak. Or, as Arkani-Hamed and his collaborators have put it: A single bar magnet is sufficient to lift a piece of iron against the downward gravitational pull of the entire earth.[2] For a more meaningful comparison, we can ask how big the mass or energy of a particle such as an electron with a unit of electric charge would have to be in order to make gravity as strong as electromagnetism. (Both mass and energy generate gravitational attraction.) The answer is the so-called *Planck mass,* which is seventeen orders of magnitude (meaning a factor of 10^{17}, or a hundred million billion) times larger than the scale where the electromagnetic and the weak force are comparable, which is already up in the hundreds of giga electron volts (GeV), about where the mass of the Higgs lies. So the question, Why is gravity so much weaker than the other forces? can be restated as a question about a difference in energy scales. In short, it is the same as the hierarchy problem: Why is the Higgs mass so small?

This problem—the question of where do the large differences between the fundamental energy scales in nature come from—had occupied Arkani-Hamed when he was working on his PhD project. Like many physicists at the time, he considered supersymmetry as a possible solution. Yet he found himself more and more often musing about possible alternatives. Something had to happen at the TeV scale that would ensure that the Higgs remained light. But what exactly was the question? At Stanford, Arkani-Hamed encountered two other restless minds: Savas Dimopoulos, who had already been a pioneer in establishing supersymmetry at the TeV scale, and Gia Dvali. The possibility the three physicists eventually came up with brought them straight into Alice's Wonderland: Gravity may not be weak at all; it may only appear weak because there are extra dimensions of space.

Extra dimensions were actually not a brand-new invention. As far back as 1919, the German mathematician Theodor Kaluza, a modest genius who supposedly spoke twenty-seven languages, wrote a letter to Einstein proposing a theory with extra dimensions as a possible way to unify gravity and electromagnetism. In his scenario electromagnetism would arise naturally as a byproduct of the extra-dimensional gravity. Einstein was at first upbeat about the "audaciousness of the thought,"[3] but as time passed his skepticism grew, and he let Kaluza's work rest for two years until he finally recommended that the Prussian Academy of Sciences publish it. In 1926 the Swedish mathematician Oskar Klein worked out the idea in more detail, but the theory turned out to be in conflict with experiment, allowing the world to forget about it for almost sixty years, until the first superstring revolution in 1984. *Superstrings* are the most promising idea for unifying all four forces, including gravity, within a single theory. In superstring theory, particles are understood as different oscillation patterns of tiny strings (Fig. 12.2). String theory was originally developed to describe strong interactions only, but John Schwarz of Caltech and Joel Scherk of Ecole Normale Superieure in Paris realized that the theory predicts the existence of closed strings that match exactly the properties of the graviton, the force carrier of gravity.[4] The paper containing their

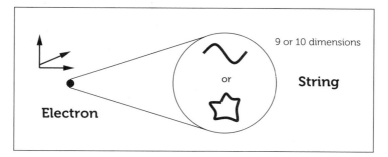

Figure 12.2. In string theory the particles arise as oscillation patterns of tiny strings.

discovery was first ignored by the community. Ten years passed before Schwarz, in another paper, this time in collaboration with Michael Green of Queen Mary College, UK, initiated the first superstring revolution and brought string theory into a leading position among candidates for a theory of quantum gravity. The problem with quantum gravity of point particles was that calculations typically yielded infinity as a result when they were extrapolated to high energies and correspondingly small distances. If particles were actually one-dimensional strings instead of point-like objects, though, a minimal distance, the string length, would be introduced into the theory, which would limit the extrapolation to arbitrarily small distances and thus would make the theory finite.

String theorists soon realized, however, that their theory worked consistently only in universes with ten dimensions, nine of space and one of time. Otherwise the calculations yielded negative probabilities and thus made no sense at all. Kaluza and Klein were suddenly back in the game!

But why are these extra dimensions not perceived, if they in fact exist? Physicists came up with three possible answers, and all of them are still being discussed, both individually and in combination. In the order of their discovery, as well as increasing degree of abstraction they are:

- Compactification
- Branes
- Warping

Compactification describes the idea that an extra dimension is so small that it can't be resolved. Just as the resolution of a microscope is limited to objects larger than the wavelength of light (this is actually the reason why X-ray microscopes are more sensitive than microscopes that use visible light), elementary particles respond only to objects larger than their wavelength. Higher energy corresponds to smaller wavelengths and thus better resolution.

But in what sense can an extra dimension actually be small? Consider the present page of this book. The page is two-dimensional; every point can be characterized by its coordinates in height and width. Now you tear the page out and roll it up tightly, starting from the edge. Once you have curled the page tightly enough, and you look on it from a distance—meaning with low resolution—the page appears one-dimensional, every point on it being characterized by a single coordinate of length, only. It's like a tightrope walker who sees her rope as a one-dimensional line, while an ant perceives the rope as a two-dimensional surface (Fig. 12.3). Now you can buy a new book.

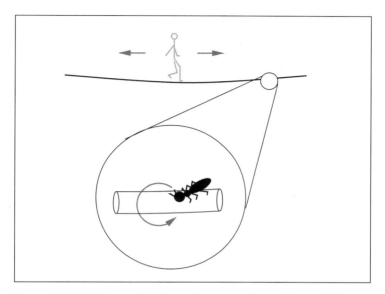

Figure 12.3. Why have extra dimensions never been observed? One possibility is that they are curled so tightly that they can't be perceived with limited resolution. It's like a rope perceived by a tightrope walker as one-dimensional but by an ant as two-dimensional.

The second idea—*branes*—was discovered in the course of the second superstring revolution. Jin Dai, Rob Leigh, and Joe Polchinski (University of Texas) as well as Petr Horava (Caltech), discovered in 1989 that string theory contains not only one-dimensional strings, but also higher-dimensional objects, so-called branes—short for membranes. These branes can be two-dimensional surfaces fluttering like sheets in the universe, or they can waft as three-dimensional volumes through space, or they can have more dimensions than we can even picture—up to nine. Branes played a crucial role in the second superstring revolution that was pushed during the mid-1990s by Princeton physicist Ed Witten and others. During this period the string theorists came to understand that the five different superstring theories discovered up to that point, as well as an eleven-dimensional supergravity theory, were only different limiting cases of one all-embracing theory, the so-called *M-theory*. To this end Ed Witten, who became the first physicist to win the Fields Medal—tantamount to a Nobel Prize for mathematics—and Joe Polchinski showed how one of the string theories behaves if the strength of the force between two interacting strings is allowed to increase. Their result was that the original theory, at large force strength, becomes equivalent to one of the other four string theories at small force strength. In the course of this process the one-dimensional strings stretch into higher-dimensional branes. In the following years, more and more such *duality relations* were found, so that the conclusion that all known theories were only different limiting cases of the all-embracing M-theory was natural. The M in M-theory refers optionally to "membrane," "mother of all theories," or to an upside-down version of W for "Witten."

As I discussed above, branes occur with all kinds of dimensions, from zero (point particles) through one (strings) up to nine, meaning that they can, for example, contain our entire universe with its three space and one time dimension. In string theory the branes constitute surfaces on which open strings end. In other words, open strings are attached to branes (Fig. 12.4).

We can now assume that our entire universe is a $3+1$-dimensional brane embedded in a higher-dimensional space. All objects in this universe are made out of particles

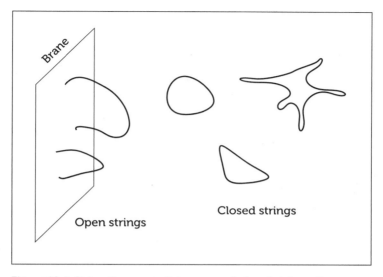

Figure 12.4. String theory predicts open and closed strings. Open strings end on branes. Particles described by open strings are thus attached to a brane.

corresponding to oscillating strings, which are—if they are open strings—affixed to branes. Apart from gravity, that is, since the force carriers of gravity—the gravitons—are represented by closed strings and thus can zoom around freely in the extra-dimensional space. In this picture of the universe, the extra dimension is not perceivable by us, since we—as conglomerates of open strings—are attached to the three-dimensional brane-surface in the extra-dimensional space and can't escape.

As attractive as string theory appears as a candidate for the unification of quantum physics and gravity, the theory has two serious flaws, which lead to heated arguments between its supporters and opponents.

First, the unification of gravity with the other forces typically happens at energies that are not accessible directly by experiment. Many physicists thus argue that string theory is not falsifiable—a basic criterion for science advocated by the philosopher of science Karl Popper. The Columbia University professor Peter Woit carried this argument to extremes when he accused string theory of being not testable and therefore "not even wrong"—a reference to an acid-tongued remark by Wolfgang Pauli.[5]

The second problem concerns the predictive power of the theory. According to present knowledge, string theory has about 10^{500} equally valid solutions, each of them describing a different kind of physics that in principle could be realized within some universe. Since no clear criterion exists on how to find the correct solution describing our universe, string theorists such as Lenny Susskind of Stanford University have evoked the so-called *anthropic principle*:[6] Since all solutions describe possible universes, all these universes may exist and we should not be surprised to find ourselves in a universe that allows for the existence of human life. In the subgroup of such universes permitting our own existence, one could then make probability statements about the fundamental laws of nature by simply counting how many of these universes respect a certain physical law.

The controversy about string theory received a new twist with the new generation of particle theorists around Arkani-Hamed, Dvali, Randall, and Sundrum (more about Randall and Sundrum below). The new blood in the particle physics elite understood string theory neither as a complete and sacrosanct theory for everything nor, on the other hand, as an unscientific, untestable speculation. Arkani-Hamed and his friends rather used string theory as a source of inspiration, as a stone quarry for new ideas that might possibly revitalize particle physics at the TeV scale. And with the help of branes, Arkani-Hamed, Dvali, and Dimopoulos (or ADD for short) realized, they could explain why gravity is so weak.

At the time that Arkani-Hamed moved across the bay, Dvali and Dimopoulos had just published a paper on how large the extra dimensions of string theory actually can be. Indeed different physicists, among them Ignatius Antoniadis, had pointed out in the past that there was no reason, really, why the extra dimensions had to be so tiny. After Arkani-Hamed and Dimopoulos had put in some work on the problem in order to understand how much sense such models actually make, they, along with Dvali when he was visiting SLAC, had a breakthrough idea. (Dvali is a first-class risk taker who is known to have snowboarded down black-diamond slopes only hours after first trying out that method of descent.)[7]

One of the facts about large extra dimensions, which was well known to other experts in this area but new to ADD, was that such theories can dilute the effect of forces: The fields responsible for forces, such as the gravitational field, can evaporate into the extra dimension and thus appear weaker in the $3 + 1$ dimensions known to us. The consequence of this thinking, for which the ADD trio became famous, was: Perhaps gravity is not weaker than the other forces at all. Perhaps gravity just appears weaker because the gravitational field is distributed not only in our $3 + 1$ dimensions but in the other dimensions as well (Fig. 12.5). A simple estimate showed how large one extra dimension actually had to be if it really was

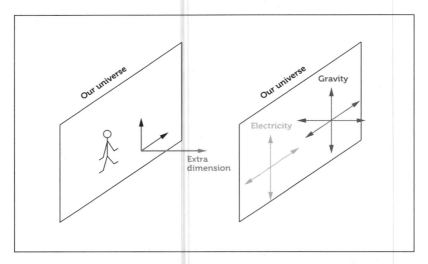

Figure 12.5. Gravity could appear weaker than it actually is as it evaporates into extra dimensions while we, the other forces, and the rest of our daily routines are confined to a 3+1-dimensional brane.

responsible for the weakness of gravity. And the answer was: huge! Up to a millimeter in size! Naturally this immediately raised the question of why such a large extra dimension had not yet been discovered. The proposal of ADD was that we were confined to a brane.

In the beginning, ADD found the idea just amusing, and they were pretty sure it had to be wrong. But after they had spent several months trying in various ways to kill it, they realized that their idea was perhaps more than an amusement; it might actually correspond to nature. In a short time they published three papers that had a significant impact on particle physics for years to come. In the first paper, ADD described the basic idea: Extra dimensions could be as large as a millimeter (in the meantime, precision experiments have determined the strength of gravity at short distances more accurately and have bounded the size of possible extra dimensions down to a tenth of a millimeter). In the second paper, together with Antoniadis, they explained how the idea could follow from string theory. And in the third paper, ADD finally derived detailed experimental bounds and predictions.

Shortly after their first groundbreaking work, ADD had realized that their idea could have fantastic consequences for future experiments. The particle physicists were captivated by these possibilities; more and more papers appeared that discussed possible implications of large extra dimensions. At small distances or correspondingly high energies, the extra dimensions supposedly could be resolved in an experiment. As soon as an experiment was sensitive enough to reach this resolution, one should notice that gravity is distributed among more than three spatial dimensions. This would imply that the evaporated gravitational field would now become observable, and gravity would become increasingly strong at short distances. Experiments measuring the gravitational force at short distances could provide evidence for this deviation. Another prediction of such models is that heavier copies of known particles exist that can propagate into the extra dimension. The wave functions of such particles can form standing waves, similar to those of an oscillating guitar string. That means, however, that their wavelengths can assume only certain values, and wavelength is related to energy. An observer who doesn't see the extra dimension finds that energy added to a particle is not entirely translated into kinetic energy (energy of motion) and so looks just like mass (intrinsic,

motion-independent energy). Since the energy of the oscillation corresponds to the oscillation patterns that fit into the
extra dimension, the impression for the observer is simply a
sequence of copies of the original particle with increasing
masses, where the smaller is the extra dimension, the larger
are the mass gaps.

It probably sounds even more astonishing that particle collisions may, as a consequence of the amplified gravitational
force at the energies reached, produce black holes (Fig. 12.6)—
bizarre objects of general relativity (Chapter 13), on whose
boundaries time is frozen to a halt. While these black holes
would immediately evaporate, a study of their decay products
could reveal groundbreaking insights about quantum gravity.
Also, other energy realms, which appeared to be out of reach
in the traditional 3 + 1-dimensional particle theories such as
the GUT scale, could now become accessible.

And finally—only three months after they had discussed
the phenomenology of their model—ADD, in collaboration
with John March-Russell (then at CERN, now at Oxford) and
Keith Dienes, Emilian Dudas, and Tony Gherghetta (then all
at CERN), concluded that extra dimensions could also shed
light on another pending problem: the question of why the
neutrino mass is so small.

At first the problem of the small neutrino mass seemed to
be made worse, since if there is no large energy or mass scale
in the theory—if GUTs, quantum gravity, and strings await
us already at the TeV scale—then there is also no large mass
scale anymore for the right-handed neutrino, which is needed
in the seesaw mechanism. Consequently, the favorite model
of neutrino theorists ceases to work. In view of this an alter-

Figure 12.6. Simulation of the signal of a black hole inside the ATLAS detector at the LHC. When the black hole evaporates into other particles, an analysis of these resulting particles could deliver totally new knowledge about quantum gravity. Also, other energy scales such as the GUT scale, which were inaccessible in 3+1 dimensions, could in extra dimensions suddenly be closer than they appear and thus be accessible to experiment. (Courtesy CERN)

native was needed, and a natural question was whether large extra dimensions, if they can explain the weakness of gravity, also provide an explanation for the smallness of neutrino mass. The idea proposed by Dienes, Dudas, Gherghetta, ADD, and March-Russell relied on the fact that string theories predict many fermions—matter particles—that bear no Standard Model charges, and, like the graviton, are described as closed strings. Among these particles are the quanta of fields that describe the radius of the extra dimension. Such particles (or, more accurately, their supersymmetric partners) have exactly the properties of a right-handed neutrino: They possess neither color nor electric charge nor isospin, which is the charge of the weak interaction. And, as closed strings, they can freely travel within the extra dimension. Now, we remember that masses correspond to transitions between left-handed and right-handed. Such a transition can happen only where the particle is located, however. In extra-dimensional models the left-handed neutrino is trapped on the brane while the right-handed neutrino can roam in the entire extra dimension. The transition between left and right is thus suppressed proportional to the small overlap of the respective wave functions for the left-handed and right-handed neutrinos (see Fig. 12.7). The consequence is that the larger the extra dimension, the smaller the neutrino mass. In the minimal model of ADD and March-Russell, the neutrinos are pure Dirac particles, while Dienes, Dudas, and Gherghetta introduced a complete seesaw mechanism, which later was thoroughly analyzed by Apostolos Pilaftsis of the University of Manchester, UK, in collaboration with Gautam Bhattacharyya and me. In this work we also explained that such an extra-dimensional seesaw mechanism can lead to large lepton-number violation and thus observable neutrinoless double beta decay, namely when the branes are shifted away from certain fixed points within the extra dimension.

On the other hand, ADD and March-Russell had also speculated that the neutrino could acquire a small Majorana mass due to the breaking of lepton-number conservation on a distant brane. If the particle transmitting the breaking of lepton-

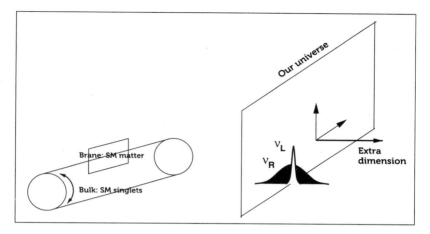

Figure 12.7. According to ADD and others the neutrino could be so lightweight because the left-handed neutrino is localized on a brane while the right-handed neutrino is spread out in the extra dimension. The transition of left-handed to right-handed, which corresponds to a mass, is thus suppressed, since the left-handed and right-handed neutrinos are rarely at the same place.

number conservation to our brane possessed a mass, then its propagation through the extra dimension would be suppressed at small energies, since it is possible only via a quantum fluctuation, very similar to the way in which the massive W and Z bosons suppress the strength of weak interactions. As lepton-number violation reaches us only weakened in this way, it can generate only a small Majorana mass. Similar ideas become even more interesting when we come now to the warped scenarios of Randall and Sundrum.

The third idea about why extra dimensions could be hidden from observations is inseparably connected with the name Lisa Randall. As a young girl Randall loved *Alice's Adventures in Wonderland,* where dimensions change in strange ways. In the summer of 1998 the ideas of ADD were a hot topic at numerous conferences, and it wasn't surprising that Randall was immediately hooked after listening to a couple of talks on extra dimensions at the SUSY conference in Oxford.[8] For the rest of the summer Randall, now a young professor at the Massachusetts Institute of Technology, stayed in Boston, where at this time her old friend Raman Sundrum was visiting. They had overlapped as postdocs at Harvard a couple of years before. Sundrum had already been working on extra dimensions, and soon the two of them were sitting together in the soda shop at the MIT student center animatedly speculating about branes and warped extra dimensions.

Warped is the crucial word here, and represents the third possibility for hiding an extra dimension. What distinguishes the Randall-Sundrum (RS) models from the ADD model is that the extra dimension in the RS models is not flat, it is *warped*. And warping means that space-time is curved in a specific way: The curvature of a $3+1$-dimensional surface parallel to our universe depends on its location in the extra dimension. Or more accurately: The farther you plunge into the extra dimension, the more strongly space and time are compressed.

Randall and Sundrum first studied models in which the extra dimension is bounded by two branes, one on each end of space-time. When they analyzed the space-time curvature of a model where the energy is positive on one brane and negative on the other, they made an interesting observation: The resulting curvature of space-time—the warping—pushes the gravitational field toward the brane with positive energy.

If one goes on to assume that the Standard Model particles and forces, meaning our entire $3+1$ universe, is located on the other brane with negative energy, the gravitational field would reach our brane considerably weakened. The calculated suppression is an exponential function of the distance of our brane from the other, positive-energy brane. Consequently even a small distance results in a huge—exponentially amplified—suppression.

In this work Randall and Sundrum obtained two results that would make them even more famous than ADD (Randall was for several years the most-cited particle theorist in the world). First, the hierarchy problem, the question of why gravity is so much weaker than the other forces, can be solved. And second, the extra dimension responsible for this effect doesn't have to be large. In this way Randall and Sundrum avoided an important criticism of the work of ADD, namely that ADD, in order to explain the extraordinary difference in strength between gravity and the other forces (the large hierarchy), had introduced a large extra dimension, and it was totally unclear how such a large extra dimension could be stabilized. From this point of view ADD had only reformulated the problem. The RS extra dimension, by contrast, can be small, and thus really solves the hierarchy problem without introducing a new hierarchy elsewhere.

Not long afterward, Randall and Sundrum made another spectacular discovery: As the gravitational field within a warped extra dimension is being pushed onto a brane, the actual size of the extra dimension becomes irrelevant. The extra dimension can be small, but it doesn't have to be. It can even be

infinitely large, meaning it can extend arbitrarily far behind our brane, without losing its attractive features. Now a
huge playground was opened: The consequences of ADD
and RS models are quantitatively different but resemble each
other qualitatively. Again black holes, Kaluza-Klein states,
and strings at the TeV scale are being predicted, in reach of
the LHC.

In addition, the RS model can also lead to small neutrino
masses, as was shown in 2000 by Yuval Grossman (then at
SLAC and now at Cornell University) and Matthias Neubert
(then at Cornell University and now at Mainz University in
Germany), as well as by Qaisar Shafi (University of Delaware)
and Stephan Huber (now at the University of Sussex in Brigh-
ton, UK). To this end, the right-handed neutrino is being localized at the distant brane and, much like the case where
lepton number was broken on a distant brane, the spread of
the wave function of the right-handed neutrino toward our
brane is suppressed, implying a small overlap of wave functions of left and right and consequently a small mass.

In the context of these ideas, the neutrino is the only known
particle that feels the extra dimension, and thus in an instant
it becomes not only a probe for GUT physics, but also for extra dimensions: Neutrinos could travel through the extra dimensions and yield information about their properties!

Not much later, Kaluza-Klein excitations of the neutrino
were discussed as candidates for the dark matter in the universe, and their influence on neutrino oscillations was analyzed, for example by Dvali and Alexei Smirnov. But as we
shall see, even stranger consequences are possible if neutrinos
really travel in extra dimensions and if these extra dimen-

sions have certain properties, such as fluctuating branes or *asymmetric warping.* In such cases, the passage of neutrinos through the extra dimension could appear as a shortcut. Neutrinos could then be faster than light, with all the mind-boggling consequences relativity has in store for us: the theory in which one can twist time.

~ 13 ~

Einstein's Heritage
What Is Time?

Actually, nobody has the faintest clue what time really is.

Why is time always running forward and never backward? Or in circles? Would backward-running time be possible at all? Is it possible to send messages, or even to travel, back in time? The reason physicists discuss the possibility of time travel at all can be traced back to the theory of general relativity.

The concept of a universal and ever-increasing time that is consistent for all observers came to a sudden end when, in 1905, a technical expert third class in a Swiss patent office used his spare time—and sometimes also his work time—to think about the universe.[1] Naturally he had to hide his notes and calculations hurriedly under his desk whenever his supervisor entered the room. Albert Einstein had graduated in 1900 with a teaching diploma from the Swiss Federal Institute of Technology, the ETH (after its German-language name), in Zurich, but due to his dubious work ethic and his notorious individualism (he was often absent from lectures), he had offended many people at the Institute. After Einstein's applications for research positions were unsuccessful, he accepted the position at the patent office, doing work that left him not only enough spare time for research but also peace and isolation from the scientific community. Eventually Einstein found the time to mull over a problem that had been haunting him since he was fifteen years old.[2] Within a single year, his annus mirabilis, 1905, Einstein laid the foundation for becoming the most famous physicist of all times, so omnipresent that my mother, when she was a twenty-year-old art student in 1961, could draw him within minutes using only a few brush strokes in a way that made him immediately recognizable (Fig. 13.1).

The puzzle that served as a starting point for Einstein was the question of what a light ray looks like to someone who is moving along with it. Light is an electromagnetic wave, and Einstein wondered why the theory of electromagnetism did not provide an answer to this question. Einstein was put on the right track by, among other things, an eighteen-year-old experiment.[3] In 1881 and 1887, Albert Abraham Michelson and Edward Morley had used a spectrometer to measure a possible difference in the speed of light parallel and perpendicular to the earth's orbit around the sun.

The experiment yielded the astounding result that the speed of light seems to be the same in all directions and regardless of the motion of the earth through a presumed background *ether*. Einstein's explanation of this and other related observations was radical, and it followed the thinking of the positivist philosopher Ernst Mach: What can't be observed does not belong in a scientific theory. If one finds that the speed of light is constant, one thus should use this as starting point for

Figure 13.1. Einstein, drawn by Frigga Päs. (Courtesy Frigga Päs)

Figure 13.2. Superhero bet. Irrespective of how fast Supergirl chases the light ray, it always escapes her with the same speed of light.

the theory instead of trying to save one's prejudices with additional assumptions.

The consequences of this simple principle remind one of a bet entered into by a couple of superheroes (Fig. 13.2). Superman has a flashlight. Supergirl has an enormous amount of energy available and is in perfect physical condition. She bets Superman that she can overtake the light coming from his flashlight. But she is up against Einstein's principle that the speed of light is the same for all observers. She can't win. When Superman switches on his flashlight, Supergirl chases the ray as fast as she can. Even if she could race through space at half of the speed of light, the light ray would still recede from her with the same speed as if she were standing still. The reason for this is that the faster Supergirl moves, the more her length scale gets shortened and the slower her clock runs. The result is that light always covers the same (now shrunken) distance in a unit of (now stretched) time. Einstein promoted the speed of light to a benchmark for the determination of time. The stretching of time can be understood if one considers a spaceship in which a certain time interval is defined as the duration a light ray takes to cover a certain distance. A certain fraction of a second, for example, has elapsed once the light ray has been emitted from its source, has hit a mirror, and is reflected back to its origin. Now we assume that the spaceship is moving (Fig. 13.3). For an observer at rest, who sees the spaceship passing by, the light ray, which now also has to keep up with the speed of the spaceship, covers obviously a larger distance. As the speed of light is always the same though, the observer sees time in the moving spaceship going by more slowly than on his wristwatch. This is indeed one of the most astonishing consequences of relativity: Time goes more slowly in moving systems. If the velocity of the spaceship were actually to reach the speed of light, time would freeze. But a massive object such as a spaceship would need infinite energy to get accelerated to the speed of light.

But that's not everything! Within the framework of relativity, two observers moving at different speeds can't even agree on the chronological order of two events: Whether one event

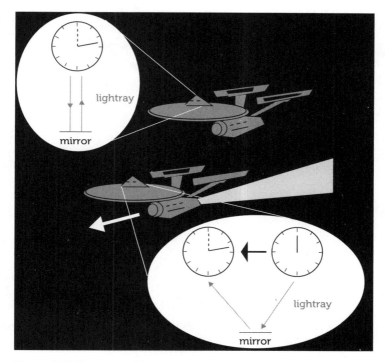

Figure 13.3. Two spaceships, one at rest, the other one moving. In both ships the time span of 12 fractions of a second is defined as the time needed for a light ray emitted from a source to hit a mirror and be reflected back. In the moving spaceship the light covers a larger distance. As the speed of light is always the same, time thus must run slower in the moving spaceship.

happens before or after another one depends on the observer! To comprehend this, imagine the following encounter in space: While you are drifting in space, your peppy friend Jane high-tails by in her spaceship (Fig. 13.4). According to relativity, as long as Jane is neither accelerating nor decelerating each of you has the right to consider yourself at rest and the other one as moving.

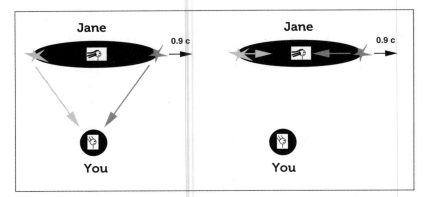

Figure 13.4. Relativity of simultaneity. You see both flashes at the same instant, meaning you conclude that they were set off at the same time. Jane, moving toward the flash at the bow, naturally sees it first and concludes that it was set off before the flash at the stern.

Now Jane sits right in the middle of her spaceship and, in the instant when she passes you, two lights flash: a green one at the bow and a red one at the stern. The red and the green light flashes reach you at the same time, so you conclude that the lights have switched on *at the same time.* Jane is moving toward the green flash, though. So she sees the green light first. But light has the same speed for anyone, irrespective of whether red or green, and whether it is Jane or you who observes it. The only possible interpretation for Jane is thus that the green light switched on *before* the red light.

This phenomenon is known as the *relativity of simultaneity* and has crucial consequences for the understanding of causality in physics: Two observers can disagree on whether two events happen simultaneously or not, and both can be right about it. If you were moving toward the red light while Jane was passing by, then you would have spotted the green flash *later than* the red flash. That means that a moving observer can perceive two events in reverse order! With this, of course, one question immediately looms large: Can an event happen before its own cause? Can beer be drunk before it was brewed? Can one become tipsy before taking the first sip? Should one decide, then, not to take that sip at all? In special relativity the answer is no, and the reason becomes clear if one considers space-time diagrams and light cones.

A space-time diagram is placed in a coordinate system with space and time axes. It can be used to study the movement through space and time (Fig. 13.5): The horizontal axis describes the position in space, while the vertical axis describes the instant in time, multiplied by the speed of light. Points above the vertical axis have positive time, meaning that this half-plane corresponds to the future. Below the horizontal axis are negative times corresponding to the past, and

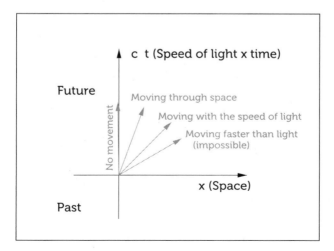

Figure 13.5. Space-time diagram.

the origin at the intersections of the axes depicts the *here-and-now*. Every point in the diagram corresponds to a specific location and a specific time, and every movement corresponds to a line or curve through space-time. If you don't move at all, for example, you still move through time. Being at rest thus corresponds to movement along the vertical axis. The faster you move through space, the less steep is your resulting movement curve within the space-time diagram. If you could race through space with the speed of light, meaning your ratio of covered distance to elapsed time would be the speed of light, this would correspond to a movement away from the vertical axis at a 45-degree angle. Light rays move along these lines. An even flatter line would correspond to superluminal motion, and thus would be impossible.

The light cone is now the hourglass-shaped figure formed by the 45-degree lines of light rays going from the here-and-now in all directions into the future and coming to the here-and-now from all directions in the past. In the cutaway diagram in the figure, there are four such lines (Fig. 13.6). All space-time points inside the future light cone are reachable in principle, if one just hurries enough. All points in the past light cone could have affected the here-and-now, since from each of these points sufficiently fast travelers or signals could have reached us without exceeding the speed of light. The region outside the light cone, however, is inaccessible for subluminal signals. It can not be reached in the future, and it couldn't have had any influence on the here-and-now from the past.

So how does the light cone explain why, in special relativity, effects always follow their respective causes, for every

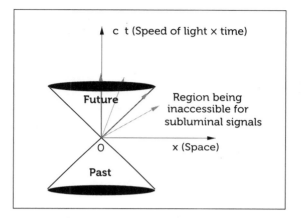

Figure 13.6. Light cone. Inside the cone resides the future, which can be affected by the here-and-now, and the past, which could have influenced the here-and-now. Events from outside the light cone neither affect the here-and-now nor are affected by it.

observer? The reason is that the sequence of the here-and-now and some arbitrary event can be perceived in reverse order by a moving observer only in the case where this event is situated outside the light cone. And events outside the light cone can neither affect the here-and-now nor be affected by the here-and-now.

In summary: Every event that is the cause of an incident in the here-and-now lies in the past or, more accurately, in the past light cone. And every event that could be caused by the here-and-now is situated in the future or, more accurately, in the future light cone. And this remains true for any here-and-now and every observer in the universe. At least in the context of special relativity.

Naturally one asks now: What if? What if you could travel faster than the speed of light? In that case you would arrive outside the light cone, and a moving observer could perceive the order of departure and arrival as reversed:[4] she would see you arrive before you departed. And this moving observer could send you, again superluminally, back to your place of origin, where you could arrive before your own departure and meet yourself as you are preparing for your trip (Fig. 13.7). In short: One could travel into the past! But luckily for causality, superluminal travel requires infinite energy, and is thus impossible in special relativity. But then again Einstein was not done at all with his disassembly of space and time.

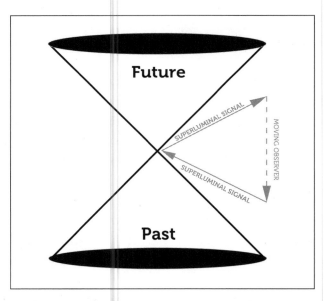

Figure 13.7. Time travel would be possible if one could move faster than the speed of light (superluminally). (1) Leave the light cone. (2) Move in such a way that departure and arrival get reversed. (3) Return superluminally to the point of origin. Result: you meet yourself before you have started your journey.

It took Einstein eleven years of hard work and many set-backs before he managed to extend his theory of special relativity to general relativity, which now also described gravity. Again Einstein followed the credo of Ernst Mach when he mulled over the question of why mass plays a role in two totally different contexts: as a gravitational mass and as an inertial mass. The inertial mass describes the property of heavy objects to resist being set in motion by an accelerating force: It is easier to push-start a pedal scooter than an SUV. And also more comfortable to be hit by a pedal scooter than by an SUV, at least at the same speed. That means that the inertial mass describes both resistance against acceleration and against deceleration. Gravitational mass is a different concept. It measures how much something weighs, or, more generally, how much it is pulled by gravity. Einstein now argued: If there is no difference between inertial and gravitational mass, then they must be the same thing—whether one carries a mass and thus counteracts gravity or accelerates it and experiences the resistance to acceleration. Or, expressed in a different way: Everything that is not free-falling is being supported somehow. We get a hint of this feeling when an elevator starts to accelerate downward. Free-falling feels like weightlessness, and according to Einstein's relativity, it is exactly the same. If you let go of an apple while you are in free-fall, you will see that the apple floats next to you. On the other hand, an elevator at rest on earth feels just like an elevator accelerated upward in remote outer space. In both cases a dropped apple will be accelerated toward the elevator's floor (Fig. 13.8).

So you can say that gravity itself is not a force, it is only a reflection of accelerating a body out of its natural state of

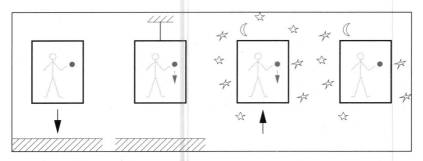

Figure 13.8. The elevator experiment of general relativity. A free-falling elevator in the gravitational field of the earth (left) appears just like one floating weightless in distant space (right), and an elevator at rest in the gravitational field of the earth (center left) appears just like one being accelerated in a weightless environment (center right). Correspondingly, a dropped apple floats (left and right) or gets accelerated toward the elevator floor (center left or center right).

free-fall. You may then well ask why the earth orbits the sun and the moon orbits earth if no force acts on them. Shouldn't they move along a straight line like a billiard ball on a flat table? Ironically they do exactly that! They follow a straight trajectory, but this trajectory leads around the sun or the earth because space itself is curved. The mass of the central body warps space in such a way that something like a trough in the billiard table is created (see Fig. 13.9).

Einstein's general relativity leads to jaw-dropping predictions such as the one about black holes, which are corpses of burned-out stars, whose mass got so compressed in their gravitational collapse that inside a certain radius (called the *Schwarzschild radius*) gravity does not allow even light to escape and time is frozen to a standstill. If an astronaut were to plunge into such a monster, on looking back he would witness the entire history of the universe unwind faster and faster, while his

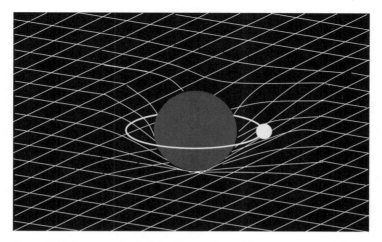

Figure 13.9. In general relativity the motion of the moon around the earth is not the consequence of an attractive gravitational force. Instead, space-time is warped by the earth, and the moon moves along a straight line in curved space. This straight line then goes around the earth.

friends left behind on a space station would see him fall more and more slowly into the black hole until his image finally reaches the Schwarzschild radius and freezes. But even if astrophysical observations provide strong hints for the existence of such timeless monsters in relativity, Einstein himself never seriously contemplated that time could ever run backward. Not, that is, until Einstein—who had fled Nazi Germany and found a new home at the Institute for Advanced Study in Princeton—became friends with one of the greatest logicians in history.

How to Build a Time Machine

Nobody would have anticipated the present that Albert Einstein received for his seventieth birthday from his friend Kurt. Least of all probably Einstein himself. Kurt Gödel was an unimpressive, small man, then in his mid-forties, lonely and pessimistic, hypochondriac and neurotic—full of fear that he might be killed by jealous colleagues or poisoned by refrigerator emissions, believing in ghosts, and barely eating anything, apart from baby food.[1] In spite of their different personalities, Gödel had for quite a while been accompanying the cheerful Einstein on his extensive walks around Princeton (Fig. 14.1). And in doing so the two men, both refugees from Nazi Germany and both enjoying the chance to converse in German, became friends. Einstein actually said that the privilege of walking home with Gödel was the only reason he went to his office at all.[2] And while Gödel tried to sell Einstein on the merits of his favorite movie, Walt Disney's *Snow White and the Seven Dwarfs*, Einstein explained to Gödel his theory of relativity.

Figure 14.1. Kurt Gödel and Albert Einstein, colleagues at the Institute for Advanced Study at Princeton. (Photo by Richard Arens, courtesy AIP Emilio Segre Visual Archives)

Gödel was no physicist, but he was curious about anything with mathematical content. As a young man he had already shaken an entire discipline of science—mathematical philosophy—to its very foundations, when he realized that every logical system, as long as it is consistent, contains unprovable truths. A trivial example is the self-referencing sentence, *This sentence is unprovable.* If the sentence were incorrect, it would be a provable incorrect statement and thus be inconsistent. With this finding, the dream of the mathematicians to find a formal axiomatic basis for logic and mathematics was rendered obsolete.

Now, eighteen years after he had found his "incompleteness theorems," Gödel made a physics discovery that in fact should have come as a bombshell in its own right. As he learned more and more about relativity, Gödel must have asked himself at some point: What is it in a theory of warped time and space that actually forbids one from traveling back in time? So he sat down, played around with the formulae, and finally came up with a startling result. The answer is: nothing. General relativity indeed has solutions that allow for time travel. And writing this up was his gift for Einstein's birthday. Specifically, it can happen when the time dimension is warped into a closed curve so that by always going forward in time one can arrive at one's initial point again.

The idea that one can travel back into the past and change it is more a perennial of popular culture than an area of scientific research: from *Star Trek* to *Harry Potter,* from *Back to the Future* to Marvel Comics. Yet physicists still don't know whether time travel is indeed possible or if not, why not.

The hypothetical universe described by Gödel's equations is neither static, like Einstein's unstable solution making use of a cosmological constant, nor expanding, like our real universe, whose geometry was found by Georges Lemaitre. Gödel's universe rotates. This rotation implies that space and time comingle: If one departs in such a universe farther and farther from the center of rotation into its outer realms, at a certain point the light cones defining future and past start to tilt over. Having reached a certain distance from the center, a traveler can then journey on a closed path, meaning that while he is always going forward in time, he arrives in his own past (see Fig. 14.2).

Gödel thus concluded that time does not exist and soon started to speculate whether ghosts could time travel. The physics community reacted by not reacting. After all, one could argue that the universe we live in is not rotating and thus doesn't match the solution found by Gödel, that Gödel's universe is only a possible but not a real world, and that time travel in the Gödel universe is nothing but a mere game.

After that, it was almost twenty-five years before the topic of time travel reappeared on the scientific agenda. This time it was in a work published by a twenty-five-year-old PhD student at the University of Maryland. Frank J. Tipler had studied the space-time around a rotating cylinder, which had been calculated initially in 1936 by the Dutch physicist Willem Jacob van Stockum (who was later killed in World War II when he was on a mission as a bomber pilot and his airplane was brought down over Normandy). Tipler concluded his 1974 article about his work with the laconic statement: "In

Figure 14.2. Causality and light cones in the Gödel universe. The farther a traveler departs from the center of rotation, the more the light cones tilt. Starting from a certain distance a closed curve from the future light cone into the past light cone becomes possible. This allows for a full orbit around the center of rotation, in which the future of the traveler is identical to his own past, while for the traveler himself time always runs in the forward direction. (Courtesy NASA/ Defense Video & Imagery Distribution System)

short, general relativity suggests that if we construct a sufficiently large rotating cylinder, we create a time machine."[3]

Again one would have assumed that the physics community would be buzzing, but Tipler's work was also largely ignored. Nine years passed before it was cited for the first time in a science journal, and until 1985 the article collected only three citations. It was in that year, however, that things began to change. It started with a phone call between old college friends.

When Kip Thorne, a professor at the California Institute of Technology, picked up the phone, it was Carl Sagan on the line, an astronomer and science fiction author, whom he knew from their undergraduate days at Cornell University: "Sorry to bother you, Kip. . . . [I'm] just finishing a novel about the human race's first contact with an extra-terrestrial civilization. . . . [I] want the science as accurate as possible. . . . [can you give me your] advice?"[4] Thorne immediately realized a problem with Sagan's plot: The astronomer Eleanor Arroway (later played by Jodie Foster in the film adaptation *Contact*), was supposed to jump into a black hole close to earth, travel through hyperspace and arrive an hour later in the vicinity of the star Vega, about twenty-six light-years away. But Thorne knew it was impossible to travel through the center of a black hole into distant parts of the universe. Not much later, during a car trip down the California coast, Thorne developed an alternative way to achieve such faster-than-light trips: a wormhole.

A wormhole is a shortcut between two distant places in the universe (Fig. 14.3). It is possible that between these two places a tubular connection exists, wherein space is compressed so strongly that the locations as seen through the tube, or through the *throat,* are only a stone's throw away. The jump into the wormhole thus allows one to reach a place far away in the universe as quick as a flash. Without difficulty Eleanor can reach the inhabitants of Vega in a finite time. But that's

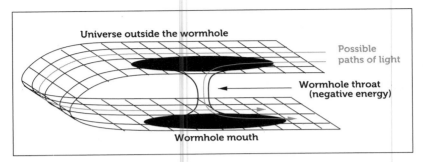

Figure 14.3. Wormhole: a shortcut through space and time.

not all: If the two places are separated far enough, or if the throat is compressed enough, a traveler jumping into the wormhole can actually be faster than light taking the normal way between the two mouths of the wormhole. This in turn means that the causality argument of special relativity doesn't apply anymore. In special relativity by itself, it is possible to alter the sequence of two events if they are located outside their respective light cones. As discussed earlier, however, a reversal of effect and cause is not possible, because the impact of an event, in order to affect another event outside the light cone, would have to be transmitted superluminally, and thus would require infinite energy.

When one goes from special relativity alone to the curved space-time of general relativity, this argument doesn't hold anymore, because a shortcut through a wormhole allows one to get from one place to another faster than light, without actually moving faster than light. Without breaking the cosmic speed limit one can jump through a wormhole and out of one's own light cone. This bewildering consequence was pointed out to Thorne's student Mike Morris a year later in a chat during a coffee break at the Texas Symposium on Relativistic Astrophysics in Chicago (one of a series that started in Texas and are now held all over the world): By using a wormhole one could travel back in time!

Thorne and Morris worked out the details together with Ulvi Yurtsever and published them in *Physical Review Letters*. One can give a simple recipe for time travel via wormhole (compare Fig. 13.7):

- Jump through a wormhole.
- Accelerate your spaceship as well as the wormhole entrance (the *mouth*) to become a moving observer.
- Jump back through the wormhole.

You would arrive before you departed!

Finally the spell was broken: Four years later Stephen Hawking wrote about the problems arising in solutions to general relativity that permit time travel, and a year after that wormholes were well known to the fifteen million US viewers of the science fiction serial *Star Trek: Deep Space Nine,* in which the Bajoran wormhole is both a sacred place and a strategically important passage. Is time travel now really possible?

What about the various paradoxes that can arise?

In the 1980s blockbuster movie *Back to the Future,* for example, Marty McFly, played by Michael J. Fox, travels back to the year 1955, where his own mother falls in love with him, disturbing the young love of his parents and endangering his own very existence. Physicists, in general less predisposed to romanticism, know this problem as the *grandfather paradox:*

A time traveler may travel into the past and kill his own grandfather, implying that he himself would never be born, meaning he could not travel back in time and kill his grandfather, which again means he was born . . . and round and round.

Another paradox, known as the *bootstrap paradox,* concerns the possibility that an event could cause its own existence in the past. A wild science fiction story by Robert Heinlein, *All You Zombies,* for example, tells of the encounter of a young man with the barkeeper of a run-down pub.[5] Although they take an immediate dislike to each other, they start talking and in the course of their conversation they find out that they are not only the same person—just on two different points of a time trip, but—again at different points—also their own father and (before having a sex change) their own mother. The end of the story hints that maybe *everybody* is the same person. Here an event is caused by nothing but itself, and the name bootstrap apparently refers to the notorious liar Baron Münchhausen, who claimed that he pulled himself out of a swamp— albeit not by his bootstraps but by his pigtail.

In view of such logical catastrophes, it is hardly surprising that both physicists and science-fiction authors regard time travel with suspicion and have thought about possible reasons why time travel may be forbidden.

In Marvel comics, for example, the Time-Keepers are a group of greenish, glowering guys with insect faces: "The Time Keepers were born at the end of time, entrusted with the safety of Time by He Who Remains. The Time Keepers were meant to watch over the space time continuum just outside of Limbo, and make sure the universe thrived."[6]

If there exists a placeholder for the Time-Keepers on earth, it is no less than Stephen W. Hawking, the ingenious former holder of the Lucasian Chair of Mathematics at the University of Cambridge, who has suffered since the age of twenty-one from the incurable neuropathy called Lou Gehrig's disease, and is probably the most famous physicist after Einstein. In a paper he published in 1992 in *Physical Review D* he announced: "It seems that there is a Chronology Protection Agency which prevents the appearance of closed timelike curves and so makes the universe safe for historians."[7] Or expressed slightly differently: There must be something that forbids time travel. In fact, Hawking had good reasons for his conclusion. If one calculates the matter distribution necessary to create the geometry of a wormhole, one encounters a curiosity: The matter distribution has negative energy. Different analyses, among others those in 2006 by Roman Buniy, my postdoc replacement at Vanderbilt University, together with Stephen Hsu (University of Oregon), have demonstrated that such spacetimes are unstable. Just as a poorly assembled IKEA shelf can collapse when a fly is passing by, your wormhole construction kit may break down if a single atom jitters wrongly.

Another argument had been advanced by Hawking himself in his Chronology Protection paper: Even if the problem with negative energies didn't exist, a particle could use such a time machine to repeatedly travel back into the past and so duplicate itself to infinity in a blink. This, however, would mean an infinitely large amount of mass or energy, which would cause the space-time to collapse. This follows for classical particles at least. Hawking then conjectured that in quantum mechanics, quantum fluctuations in empty space could build up to infinite energy. But whether this result also follows in a full-fledged quantum calculation or whether wave functions get stretched so strongly that the energy becomes diluted is not yet clear. To answer this question conclusively, most likely a quantum theory of gravitation will be necessary.

In short: It looked bad for the realization of time travel. There were encouraging voices, though. Googling the search word *wormholes*, Tom Weiler came across the comment of an anonymous Web user: "Too pessimistic, If worms can do it, people can learn how!" In any case: In the near future, time travel for people is a problem that is too ambitious. If a time machine is ever to be realized—regardless of what kind—the first passengers for sure will be elementary particles. And a possible way to realize time travel, even in view of the problems discussed, was actually acknowledged by Hawking himself: "According to string theory . . . space-time ought to have ten dimensions. . . . What this would give rise to, we don't yet know. But it opens exciting possibilities. . . . Science fiction fans need not lose heart. There's hope in string theory."[8]

15

Against Hawking and the Timekeepers

I had my first postdoc job in Nashville, Tennessee: Music City, USA. It is one of the fixed points of the Memphis–New Orleans–Nashville triangle, where almost everything that America spawned in blues, jazz, country, folk, and rock originated. If you turn on the radio in the South, almost every third song tells about one of these cities, the Mississippi River, the Tennessee River, the wide plains of Alabama, or the highways in between. Nashville is a prime destination for hopeful singers, songwriters, and bands of every kind, who play for tips in bars and restaurants and hope to be discovered accidentally by one of the big studio owners munching pizza there. As my postdoc successor, the devoted AC/DC fan Sergio Palomares-Ruiz has told me, Nashville even has a specialized pawn shop where unsuccessful musicians bring their instruments: a museum of broken dreams. And rather like the musicians, as a postdoc you dream of a permanent job at one of the big universities or research centers in Europe or the United States, when—after a long day of shuffling mathematical symbols around in the red brick temples of the time-honored Vanderbilt University—you stroll around Nashville's streets at night. And of course you dream of the single, crazy idea that could make you famous. At that time I worked with supervisors Tom Weiler and Tom Kephart on different projects concerned with the absolute mass determination of neutrinos, neutrino oscillations into sterile neutrinos, and possible derivations of the Standard Model from superstring-inspired models. But just before the end of my year in Nashville, I attended a seminar on wormholes by Tom Weiler's PhD student Liguo Song. Among other things, Liguo explained in his seminar how a wormhole could be transformed into a time machine, and he recommended to me some literature on the topic.

A few weeks later I was back in Germany, where I planned to work at the University of Würzburg on my *Habilitation*—a German degree after the PhD that qualifies a postdoc to apply for faculty positions. Shortly after getting back, I visited a European conference, Neutrinos in the Universe, in the Bavarian region called Lenggries. During the year I had spent in the United States, the idea of large extra dimensions had become a raving success, and many of the talks dealt in one way or another with this topic. During a rather boring talk I let my mind wander, and I started asking myself whether these extra dimensions couldn't contain shortcuts permitting one to go faster from one place to another than is allowed for travel constrained to our brane. Could these extra dimensions then act as a wormhole? Would even time travel through extra dimensions be a possibility? This was definitely a crazy idea. Whether it would make me rich and famous remained to be seen.

First I tried to convince some colleagues of the merit of the idea—with rather limited success. Alexei Smirnov, the father of the MSW effect (see Chapter 9), looked at me as if I had lost my marbles and asked me for any reason why neutrinos should travel back in time. Michael Kachelriess, an old friend from the Italian Gran Sasso lab who was now a postdoc at CERN, at least advised me of the works of cosmologists who had tried to solve the horizon problem with shortcuts in extra dimensions. But even Tom Kephart, my ex-boss from Nashville, who is normally enthusiastic about even the craziest ideas, thought that this proposal was a little too wild. Obviously I was not very convincing, which may have been related to the fact that my knowledge of general relativity was rather marginal, and that I couldn't come up with a good reason why neutrinos, after setting off into the extra dimension, should actually come back to the brane. So the idea was put on ice, until almost three years later when I met Tom Weiler at CERN—CERN, the European center for particle physics; the place where Rubbia's UA1 waged its epic battle with UA2 for the holy grail, the experimental confirmation of the Standard Model; the place where the World Wide Web was born; the location of the biggest and most energetic accelerator in the world; the lab that inspired this description in Dan Brown's bestselling mystery thriller *Angels and Demons:* "Looming before them was a rectangular, ultra-modern structure of glass and steel. . . . The glass cathedral . . . Quarks and Mesons? No border control? Mach 5 jets? Who the hell ARE these guys?"[1]

The reality is less glamorous, though. Most of the place resembles the scattered trailers and prefabs of an emergency accommodation, with bleak linoleum covering the floors. Yet hidden behind shabby wooden doors, in between towering piles of papers and books, work some of the smartest minds in the universe. That is, if they are not just sitting in the cafeteria, open for seventeen hours every day of the week, consuming coffee, beer, or wine out of never-dwindling taps, and discussing the most recent facts and rumors of the scientific community.

When I met my former boss Tom Weiler again there in the early summer of 2004, for the first time after almost two years, he first provided me with a detailed account of a sex show he had attended on the Reeperbahn, the famous red-light district close to the harbor in Hamburg, Germany, during a visit at the research center DESY. After the story was told in all its juicy details, Tom finally told me that just before his departure for Europe he had visited the University of Hawai'i. There, he and Sandip Pakvasa—who would be my next supervisor in the coming fall—had revived my old idea: "Heinrich, Sandip and I have started to put your idea about neutrinos taking shortcuts in extra dimensions into a somewhat more solid context. We were wondering whether such shortcuts might affect neutrino oscillations." Almost at once, Tom and I, along with Gautam Bhattacharyya (an old friend from Kolkata, who was also visiting CERN), explored this question, and what we found out was definitely enthralling: Shortcuts in extra dimensions would indeed affect neutrino oscillations, and it looked quite possible that this effect could make it possible to explain not only the oscillations of solar and atmospheric neutrinos but also the result of the LSND ex-

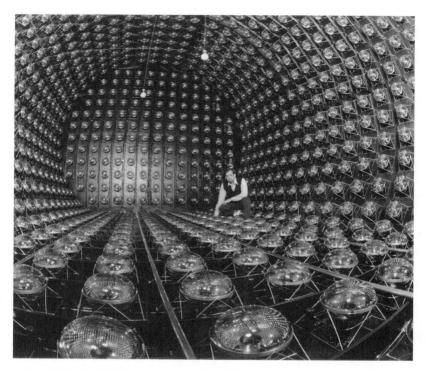

Figure 15.1. The LSND-experiment at the U.S. nuclear research center in Los Alamos, New Mexico, where during World War II the first nuclear weapons were developed. In 1996 LSND reported evidence for neutrino oscillations that still remains unexplained. (Courtesy William C. Louis, LSND Collaboration)

periment (Fig. 15.1). This LSND result, dating from 1996, provided a hint of neutrino oscillations, but it was less clear than the other neutrino-oscillation results, and even seemed to be incompatible with the other experiments.

The problem was the following: If there are three different pieces of evidence for neutrino oscillations (from solar neutrinos, atmospheric neutrinos, and LSND), then there must also be three different mass differences. Three different mass differences means four neutrinos (as long as one is not the sum of the other two, and that is not the case here). On the other hand, it is known from its half-life that the Z boson can decay into only three different neutrino families: If there were four of them, the decay would be faster than observed. The fourth LSND neutrino, if it exists, therefore can't interact with the Z boson. But since the weak interaction is the only force neutrinos feel, this hypothesized fourth neutrino doesn't interact at all (apart from a possible interaction involving the exchange of Higgs bosons). Such a neutrino is called *sterile,* but as we have seen before, one usually assumes that such neutrinos are extremely heavy, typically of the order of the GUT energy.

But that was not enough: Even if one accepts the existence of such a light sterile neutrino, one has to mix it with the known neutrinos in order to explain LSND. What LSND had reported were transitions of muon antineutrinos into electron antineutrinos. If such an oscillation were triggered by sterile neutrinos, it would imply that both electron and muon antineutrinos were mixed with the sterile neutrino. The consequence would be that both electron and muon neutrinos could oscillate into sterile neutrinos, or disappear—which is the same thing, since sterile neutrinos cannot be detected. In order to explain LSND, either the mixing of electron neutrinos had to be large enough that one should see reactor neutrinos oscillate into sterile ones, or the mixing with muon neutrinos had to be large enough to make accelerator neutrinos oscillate into sterile ones. None of this had been seen.

Now of course one can take the position that LSND is wrong. Or one can wonder whether LSND may be explained by a nonstandard oscillation mechanism. If there are sterile neutrinos and if extra dimensions exist, then it isn't unlikely that sterile neutrinos can propagate in these extra dimensions. Moreover, these extra dimensions may be warped or compressed in a way that allows the sterile neutrinos to be faster there than their non-sterile siblings on the brane.

If this were correct, then this mechanism could provide an explanation for LSND that resembles the MSW effect for solar neutrinos.

As in the MSW picture, every neutrino is a mixture of different masses. While in the sun's interior the interactions with matter slow down the faster (meaning lighter) neutrino components (compare Chapter 9) and in this way affect the neutrino mixing, in the brane picture the heavier and thus otherwise slower components could actually be faster by taking a shortcut through the extra dimension. Just as in the case of the MSW effect, this process can amplify the mixing when it cancels the mass differences and suppress the mixing when it enhances the mass differences. And since this effect is energy-dependent, the mixing can be larger at small energies than at large energies. And finally—since reactor experiments, LSND, and accelerator experiments are all conducted at different energies—the oscillations of accelerator experiments may be suppressed and the problem of finding oscillating LSND neutrinos and non-oscillating accelerator neutrinos vanishes into thin air.

Even problems in cosmology may be avoided in this scenario, if in the hot, dense plasma of the early universe neutrinos were kicked out more often than today into the extra dimension due to brane fluctuations, or scattering. The shortcuts of sterile neutrinos thus became more efficient and the mixing would be suppressed at even smaller energies. In this way the sterile neutrinos wouldn't be produced from neutrino oscillations and couldn't do any harm, for example to nucleosynthesis (see Chapter 10).

A few months after I had left CERN and moved to Hawai'i, Sandip Pakvasa and I came back to this point. And this was the state of the art on that afternoon when I was wandering around the Manoa campus and staring at the cliffs. Somehow I couldn't forget about the notion of neutrino time travel. But how could one bend the brane in a way that would make it look like a wormhole? On that afternoon I finally realized that this wouldn't be necessary at all. If the brane underwent tiny oscillations, or if the extra dimension parallel to the brane were compressed, then a sterile neutrino would naturally be faster in the extra dimension than in the brane. In that case, at least so it seemed to me, the extra dimension should behave like a wormhole. All that was left to do therefore was to construct an extra-dimensional brane-world that had these properties, and then check whether the energy-momentum distribution would be as discouraging as for wormholes in $3+1$ dimensions. There was a catch, though: As I mentioned already, I was not very well versed in general relativity. So what do you do if you are a neutrino physicist and you think you just invented a time machine? Well, you can follow three easy steps:

1. Log yourself into www.amazon.com.
2. Buy the excellent textbook *Lorentzian Wormholes,* by Matt Visser.
3. Ignore all insults of the type "Psycho-ceramics warning: Crackpots are politely requested to refrain from reading this paragraph"[2] (Sorry, Matt . . .).

In parallel I rummaged through the appropriate literature, and infected first Sandip and later also Tom Weiler with my enthusiasm. Pretty soon I realized, though, that I was too optimistic with respect to oscillating branes. If the brane is bent in the extra dimension or flapping like a towel in the wind, then the path a light ray takes on the brane is longer than the direct way, instead of being shorter inside the extra dimension. Instead of an extra-dimensional shortcut we have a brane detour. But something still seemed to hold promise: a scenario that refined the warping idea of Randall and Sundrum, developed in 1999–2000 by Dan Chung and Katie Freese, as well as by Csaba Csaki, Josh Erlich, and Christophe Grojean. In such a scenario space is warped, but time isn't. Consequently the space-time parallel to the brane is compressed, and the more strongly so the deeper one plunges into the extra dimension (Fig. 15.2). This is known as *asymmetric warping.* Originally such space-times were proposed in order to solve the horizon problem (the extremely homogeneous matter distribution in the universe) as an alternative to inflation, or to dilute the cosmological vacuum energy to its tiny actual value. In such space-times a simple analogy with wormholes

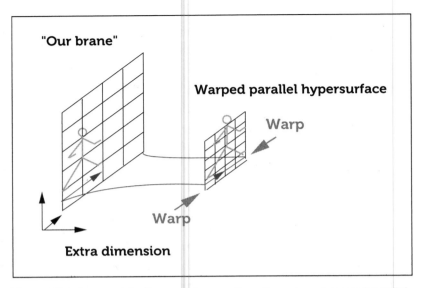

Figure 15.2. Asymmetrically warped space-time. Space is compressed parallel to the brane—and increasingly strongly the farther one moves away from the brane.

could be realized. So here is a recipe for time travel using extra dimensions:

1. Take an extra-dimensional shortcut.
2. Accelerate your position to become a moving observer.
3. Take the extra-dimensional shortcut back and arrive before you departed.

Indeed my calculation seemed to confirm that the path of a trip back in time, or a closed time-like curve, can be constructed explicitly. Quite a few colleagues, among them many who had been working on general relativity for a long time, remained skeptical. In particular Dan Chung and Josh Erlich warned us that there had to be an error in our calculation. At first we believed we could ignore this criticism, but after both the discoverer of inflation, Alan Guth, and the referee of our paper at *Physical Review D* produced the same arguments, we started to feel insecure. I checked my calculation and indeed found a sign error whose correction implied that the way back after the acceleration of the observer wasn't accessible anymore. The time trip couldn't be accomplished; the timelike curve couldn't be closed. At this point things started to become embarrassing, as several popular-science and science-fiction magazines had reported our result already. Then, as if it couldn't get worse, Lenny Susskind, one of the fathers of string theory and many other important developments in particle physics and cosmology, put out a paper entitled "Wormholes and Time Travel? Not Likely," which clearly contradicted our understanding of the matter. Fortunately Susskind backpedaled only a month later, putting out a sequel in which he criticized his own previous paper humorously by stating "the author of that paper demonstrated that he didn't know what he was talking about" and "the argument which reveals the author's deeply held prejudices against this interesting subject is incorrect."[3]

After several tough weeks, the crucial idea on how to rescue our project occurred to me when I was running at Waikiki Beach: If you have two asymmetrically warped extra dimensions instead of just one, which in addition move relatively to each other, then one could travel forth through one of the dimensions and back through the other. The whole thing didn't seem to be too unrealistic: Once extra dimensions existed at all it was likely that there would be more than one, and in addition there seemed to be no good reason why they should all be at rest with respect to each other. We wrote down the relativistic formula and voila—it looked very similar to the time travel formulas of Gödel and Tipler. Thus we eventually had demonstrated that time travel was possible in asymmetrically warped space-times. A last question remained, and this was the question of why neutrinos, after they left the brane and set off into the extra dimension, should ever return to the brane. This problem was solved by us in 2009, when my postdoc Octavian Micu and my PhD student Sebastian Hollenberg (I myself was a respectable professor by then), in collaboration with Tom Weiler, calculated the paths neutrinos take in asymmetrically warped space-times and concluded that the sterile neutrinos actually oscillate around the brane. Therefore neutrino beams could really be used as a tool to search for time loops. In order to force them to take the shortcut within the correct extra dimension one could send them off in the resting frame and shoot them back in a moving frame, resulting in a reversal of the times of departure and arrival.

And even better, neutrino time travel in extra dimensions actually has several benefits compared with time travel via wormholes:

- As Matt Visser had remarked,[4] the recipe for how to transform a wormhole reminds one a little of the recipe for dragon stew: Step 1: Find a dragon. While a wormhole could be difficult to get hold of, the extra dimension is everywhere—in case it does exist—"just around the corner."

- Negative energy would be necessary only inside the extra dimension and only in a very mild form. (Negative energy leads to violations of the so-called *energy conditions,* which define a relation between energy and pressure. A "mild form" of negative energy is one where only the more stringent of these conditions are violated.)
- Quantum fluctuations escalating to infinity are less likely in space-times with extra dimensions.
- An experimental test would actually be feasible.

In order to actually build such a time machine, one has to be able to turn normal neutrinos into sterile ones and back, so that a detector can measure and time them. Given all the necessary conditions listed above, a technical realization would require an efficient transformation mechanism between active and sterile neutrinos, as well as a large detection probability of the elusive particles.

To transform the neutrino flavors one could then make use of the MSW effect: A slowly changing matter density could compensate the mass difference of the neutrinos so that they would be converted resonantly at a certain energy and matter density.

Marcus Chown, author of the book *The Universe Next Door,* described such a neutrino time machine in the magazine *BBC Focus:* "At a laboratory at the North Pole, a physicist flips a switch. A pencil thin beam of subatomic particles known as neutrinos stabs down through the rock of the Earth. At a laboratory on the equator a spike on a computer screen registers the neutrinos' arrival. The physicists crowding around the computer monitor are stunned. They check and re-check the timing. It cannot be true—but it is. The neutrinos arrived before they set off. Surely it isn't possible for anything to travel into the past, as these neutrinos have done? You might think so. However a team of physicists in the US maintains it is easy to envisage scenarios in which particles travel back in time—if, as many believe, we live in a universe of more than four spacetime dimensions."[5]

A first step toward such a development would definitely be to verify some of the speculative prerequisites, such as extra dimensions, asymmetric warping, or sterile neutrinos. For example, it would be interesting to see whether the presently running experiments at the LHC at CERN could reveal any evidence for extra dimensions, such as black-hole production or Kaluza-Klein excitations. In parallel, neutrino experiments search for the existence of sterile neutrinos. If these searches turn out to be successful one could, perhaps in fifty years, start to investigate the causal structure of extra dimensions with neutrinos. Whether one really finds time loops then remains to be seen. In particular, the MiniBooNE experiment at Fermilab near Chicago has tried for several years to confirm or refute the LSND results. The operating energy at Mini-BooNE is a little higher than at LSND, though, and the experiment initially used a neutrino beam and not, as LSND did, an antineutrino beam. The first results of the MiniBooNE collaboration, released on April 11, 2007, in a crowded seminar room at Fermilab and simultaneously aired via live stream over the Internet, nevertheless looked quite promising for extra-dimensional shortcuts: The data initially didn't show oscillation in the high-energy part of the spectrum, but did show a strong, resonance-like peak in the low-energy part of the spectrum, just as would be expected from an energy-dependent, resonant conversion in extra-dimensional neutrinos. The first comments were correspondingly enthusiastic: "It is indeed startling to see how well your model appears to fit our excess of low-energy events!?!" (Bill Louis, Los Alamos, LSND and MiniBooNE spokesperson); "We knew you would be happy!" (Janet Conrad, Columbia University, MiniBooNE spokesperson); "The first thing I thought about when seeing the low-energy stuff was your paper" (Ion Stancu, University of Alabama, LSND and MiniBooNE collaborations); "Wow! Your figure 5 is pretty spooky when compared to the Mini-BooNE data! Maybe MiniBooNE has discovered extra dimensions!" (JoAnn Hewett, SLAC Theory Group).

Meanwhile a more accurate analysis of the MiniBooNE data yielded a width of the oscillation events that was too narrow to be explained by shortcuts in a simple way. It is thus still unclear whether shortcuts in extra dimensions can explain either the LSND or the MiniBooNE result.

An exciting result making headlines around the world in the fall of 2011, when the OPERA long baseline experiment measured a superluminal neutrino speed for neutrinos being shot from CERN to the Gran Sasso laboratory in Italy, turned out to be wrong—the result of a loose cable. This may sound funny, but one must remember that modern particle physics experiments are so complicated and fine-tuned that an error is always quite possible.

Nevertheless the possibility of extra-dimensional shortcuts and neutrino time travel inspired the imaginations of professional writers: For example the science-fiction author Joe

Haldeman, a winner of the Nebula Award, refers in the appendix of his novel *Accidental Time Machine* to our work on extra-dimensional time travel and concludes with the words: "If you only flap your arms hard enough you are able to fly!"[6] And a contributing editor at *Scientific American,* Mark Alpert, in his first novel, *Final Theory,* constructs out of neutrino shortcuts the concept of a futuristic weapon and comes up with the plot of a slam-bam science thriller.[7]

But what about the paradoxes? They are serious. Time trips of elementary particles can lead to catastrophes such as the grandfather or the bootstrap paradox, if one can use them to transmit information into the past: You could, for example, after you have totaled your brand-new car, send yourself a neutrino message telling yourself that you should not start the trip. As a consequence, you don't depart, there is no accident, no message is sent, you receive no message, you depart, you have an accident and on and on. There do exist ideas, however, on how such paradoxes could be resolved. The Oxford physicist and quantum computer pioneer David Deutsch, for example, has speculated that consistent time travel should be possible between the different parallel universes arising in Everett's many-worlds interpretation of quantum mechanics.[8] In that case, departure and accident would happen in one parallel universe, and the receipt of the message in another, thus avoiding a direct contradiction. Nevertheless, such paradoxes are the main reason most physicists believe that some mechanism must exist to prevent time travel. The investigation of such a mechanism would give rise to significant progress in our understanding of the nature of time. On the other hand, physics has surprised us in the past with copious apparent paradoxes: Simultaneity is subjective, strict causality does not hold, and clocks run more slowly once you are moving. It is not at all impossible that the physics of time travel is only temporarily baffling. All that can be said at the present time is that neutrinos are hot candidates for the discovery of any such phenomena.

In the event that neutrino time travel is possible, and if David Deutsch is right with his many-worlds interpretation of time travel, then without doubt we live in a universe far more amazing than all the marvelous things we have learned already. The parallel universes of quantum physics would form a multidimensional web of events and their alternatives, a timeless entity, at all times connected by neutrinos, which would constitute the cement between worlds and times. Here the circle closes, and we are back at the beginning of our trip, the Whole, the One of Parmenides, in which everything that is, everything that was, and everything that could be belong together and are deeply related.

16

Into the Wilderness of the Terascale

Starting in the spring of 2010, the LHC—the largest machine ever built by mankind—produced particle collisions with 7 TeV center-of-mass energy, on its way to its design energy of 14 TeV (see Fig. 16.1). Just like the pioneers of the American west, we enter an unknown territory: a wilderness where we don't know what to expect, but where we hope to encounter new exciting physics. The hierarchy problem and the need for dark matter in the universe provide a strong motivation for these hopes.

But is there any connection with neutrinos? There are actually good reasons for such a connection, and they follow directly from a 700-year-old principle known as Occam's Razor: "Entia non sunt multiplicanda praeter necessitatem," or "Do not add new things without necessity."

With this statement, William of Occam, a fourteenth-century English Franciscan monk who was excommunicated by the pope, wanted to make the point that among different possible explanations the simplest one is the best one—that the superior theory is the one that needs the least number of concepts to explain the subject matter at hand. This principle continues to be cited in science, and it has a central significance for the role of neutrino physics at the LHC. On the one hand, neutrinos so far provide the only hint for new physics beyond the Standard Model. And on the other hand, we expect to discover new physics beyond the Standard Model at the LHC. According to Occam, it is thus not unnatural to assume that the new physics related to neutrinos and the new physics at the LHC are related.

Another hint for such a link follows from a theorem proven in 2006 by Martin Hirsch of the University of Valencia, Spain, along with Sergey Kovalenko and Ivan Schmidt of the University Federico Santa Maria in Valparaiso in Chile: The existence of Majorana neutrinos and the occurrence of neutrinoless double beta decay automatically imply the existence of other lepton-number-violating processes, for example the production of two leptons with the same charge at the LHC, the so-called *like sign di-lepton signal*. Even if the exact rate of such processes can't be predicted, their mere existence is definitely a heartening fact. And there are more concrete ideas on how neutrino physics could reveal itself at the LHC: One can summarize these possibilities by defining three frontiers of knowledge that can be explored at the LHC.

The Unification Frontier

Typically the energy scales where a GUT theory or even a version of quantum gravity such as string theory is expected to

Figure 16.1. LHC detector ATLAS, before it was closed around the collision point. A sky-high engineering marvel, 45 meters long, 22 meters high, weighing 7,000 tons, dedicated to the search for Higgs, SUSY, extra dimensions and . . . the unexpected. (Courtesy CERN)

crop up are in the range of 10^{16} to 10^{19} GeV, that is thirteen to sixteen orders of magnitude above the TeV or terascale now being investigated at the LHC. In certain models with extra dimensions, or in string theories where different theory versions with different numbers of space dimensions can be related via so-called *duality relations,* however, the energy scale can be much lower. If the mechanism of neutrino mass generation is linked to the physics at such a "low" energy scale, it should be possible to probe it directly at the LHC.

But even in theories where the unification is realized at a large energy scale, the origin of neutrino masses can be at the TeV scale. This shows up in models where the masses of the right-handed neutrinos are generated via loop quantum fluctuations into two other particles. In SUSY models, such quantum fluctuations cancel above the energy scale where SUSY particles can be produced—similar to the way in which the cancellation of the contributions to the Higgs mass solves the hierarchy problem. Such theories thus need a mechanism for neutrino mass generation at or below the mass scale of SUSY particles, such as the inverse seesaw. Models of this kind have been investigated recently by Sören Wiesenfeldt, my PhD student Christophe Cauet, and me.

The Majorana Frontier

Is the neutrino a Majorana particle after all? The best possibility for finding out is in the search for neutrinoless double beta decay. As soon as one measures neutrinoless double beta decay, one can, according to the Schechter-Valle theorem, conclude that neutrinos are definitely Majorana particles. What can't be concluded, though, is what exact kind of lepton-number-violating physics has triggered the decay, and how it is related to the mechanism of neutrino mass generation. One possibility for obtaining this information could be to measure neutrinoless double beta decay in different isotopes, as has been proposed by Frank Deppisch and me and, shortly after, by Steve Elliott and Victor Gehman of the Los Alamos National Lab. But this approach is extremely intricate experimentally, and promising in only very few cases. A better method, probably, is to search for the corresponding *like sign di-lepton decay* at LHC, which arises in various models, and to compare it with a possible signal for neutrinoless double beta decay. Ben Allanach, his PhD student Steve Kom of Cambridge University, and I have studied such signals in R-parity-violating SUSY and found very promising results. As of this writing, I am engaged with Martin Hirsch, Sergey Kovalenko, and his PhD student Juan Carlos Helo Herrera in a more general, model-independent analysis.

The SUSY Frontier

If supersymmetry exists at the TeV scale, some of the neutrino properties, such as lepton-flavor violation and lepton-number violation, would be transmitted onto the SUSY particles. The exact details of the transmission, such as quantum fluctuations of SUSY particles into neutrinos and sneutrinos via lepton-flavor-violating couplings in the seesaw mechanism, again depend on the concrete mechanism of neutrino mass

generation. SUSY particles would then undergo lepton-flavor-
or lepton-number-violating processes, which do not arise in
the Standard Model and could shed light on the mechanism
of neutrino mass generation. Such decays are studied, for example, by Werner Porod and Andreas Redelbach in Würzburg, Germany; Martin Hirsch in Valencia, Spain; Frank Deppisch in Manchester, UK; and their respective working
groups. In addition, the same interactions could reveal themselves in the search for lepton-flavor violating decays of charged
leptons, such as that of a muon into an electron and a photon
being searched for in the MEG experiment at the Paul Scher-
rer Institute, Switzerland.

A relationship between neutrino physics and supersymmetry is also suggested by the lepton number and R-parity
violating couplings that arise naturally in SUSY models and
make it possible to generate neutrino masses via quantum
fluctuations. Such models and their phenomenology have
been extensively studied for years now in the Valencia group
around Jose Valle and Martin Hirsch. Moreover, such models
also have interesting consequences for the next point, the flavor frontier, which Gautam Bhattacharyya, Daniel Pidt, and I
have been studying.

The Flavor Frontier

One of the most interesting and least understood aspects of
neutrino physics is the question of why neutrinos exhibit such
large mixing among different flavors as compared with the

rather small mixing of quarks. This question is part of the *flavor puzzle* addressing the relations among the three parti- cle families and their respective members; many groups around the world work on models that explain these relations with the help of symmetries. In particular the symmetry of the tetrahedron, proposed in 2001 by Ernest Ma at the University of California, Riverside, as well as the symmetry of the equi- lateral triangle, could play an interesting role here. My PhD student Philipp Leser, who has studied a model of this kind proposed by Ma, Michele Frigerio, and Shao-Long Chen, found that in such models the neutrino mixing can induce characteristic, lepton-flavor- and quark-flavor-violating cou- plings to the Higgs particle, which lead to exotic Higgs decays observable at the LHC. If such a signal could shed light on the solution of the flavor puzzle, this would definitely be one of the most spectacular discoveries at the LHC.

The Extra-Dimensions Frontier

Finally, there are searches for extra dimensions at the LHC. Such a discovery would definitely be highly relevant for neu- trino physics. If researchers at the LHC find extra dimen- sions, for example by detecting microscopic small black holes or Kaluza-Klein excitations, then extra dimensions almost necessarily have to play a role in the process of neu- trino mass generation. If it finally turns out that neutrino masses indeed are small because the right-handed neutrinos propagate in extra dimensions, then the step to shortcuts in extra dimensions and neutrino time travel isn't that far off anymore.

The discovery potential of the LHC alone will make the next decade one of the most exciting eras in the history of particle physics, with implications for neutrinos. But in parallel there are exciting projects upcoming in neutrino physics itself: First and foremost, experimenters will put an effort into the determination of the so-far unknown observables in the neutrino sector. Among them are the masses of neutrinos and the question of whether the neutrino is a Majorana particle. Exciting news in this respect can be expected by new double beta decay experiments such as EXO-200 in New Mexico, KamLAND-Zen in Japan, and GERDA in Italy, which, as of this writing, are taking data and have reported first results. Another experiment in Italy, CUORE, is scheduled to start taking data in 2014. Complementary information will be obtained at the tritium beta experiment KATRIN, and from cosmology.

The smallest mixing angle, called θ_{13}, among the parameters that describe neutrino mixing has recently been determined in reactor experiments—a promising neutrino source, since Reines and Cowan used reactors to discover neutrinos in the first place. A typical reactor emits 10^{20} antineutrinos per second. Until recently, the most sensitive test for θ_{13} had been provided by the CHOOZ experiment in the French Ardennes. A problem with this measurement was, however, that the neutrino flux at the source could be determined only from the heat emission of the nuclear power plant, a method that suffers from large uncertainties. And if one has only a vague knowledge about how many neutrinos are being emitted, it is rather complicated to conclude that one is missing some. The new generation of experiments, a successor of CHOOZ called Double-CHOOZ, DAYA BAY in China, and RENO in Korea, incorporate both near and far detectors, where the near detectors determine the original source flux at a distance of a few hundred meters, while the far detectors measure the flux after the neutrinos have propagated and oscillated a distance of about a kilometer. In March 2012, the DAYA BAY experiment actually reported a measurement of θ_{13}, which confirmed previous hints from other experiments and has since been further confirmed by RENO and Double-CHOOZ. The fact that the actual value of the small mixing lies close to its upper bound raises hopes that other unknown fundamental neutrino properties can be measured soon as well.

These properties include the question of whether the electron neutrino is composed mainly of the heavier or the lighter neutrino masses (known as *inverse* or *normal mass hierarchy*), as well as the question of whether the symmetry between neutrinos and antineutrinos of the opposite helicities, the so-called *CP-symmetry,* is violated. The answers to these questions are important for the flavor structure and thus for a solution of the flavor puzzle, and a new generation of experiments improves on the already crazy-appearing efforts firing neutrino beams hundreds of kilometers through the earth. Among these ambitious projects is the T2K project, with its 295-kilometer neutrino beam. There, neutrinos are shot into the earth from the J-Parc accelerator complex in Tokai north of Tokyo with a hundred times greater intensity than in the previous experiment, K2K, which confirmed the atmospheric neutrino oscillations. Super-Kamiokande saw the first T2K neutrino on February 24, 2010. While T2K probably has the best chances of observing CP violation among neutrinos, also under construction in the United States is the competing NOVA experiment, with a neutrino beam originating at Fermilab close to Chicago and aimed at a 15,000-ton detector in Ash River, Minnesota—812 kilometers away. Even farther in the future one could have superbeams (neutrinos from conventional pion-decay sources but with optimized intensity), beta beams (neutrinos from the decay of highly accelerated radioactive ions), and neutrino factories (neutrinos from the decay of highly accelerated muons) as neutrino sources. In addition, a new anomaly has appeared recently in the reanalysis of the combined data of various older reactor experiments: They seem to indicate the existence of a fourth, sterile neutrino (compare the discussion in Chapter 15). While it is not clear at all right now whether this result will hold, it could have crucial impact on the attempts to understand the results of LSND and MiniBooNE, if they are not due to faulty measurements. A direct test of LSND, which would also be valid in models with neutrino shortcuts in extra dimensions, needs to be executed at comparable energies and detector distances ("baselines") from the source. Such experiments had been proposed for the Spallation Neutron Source in the US nuclear research center in Oak Ridge, Tennessee, as well as for Fermilab, but didn't meet the test of convincing the relevant funding bodies. Also interesting in this context are measurements that deviate from LSND in energy and baseline but correspond to the situation at LSND if one looks at a combination of these quantities, such as their product. For neutrino short-cuts in extra dimensions, measurements at the reactor experiments Double-CHOOZ, RENO, and DAYA BAY could possibly deliver exciting results.

Vigorous work is also proceeding regarding the cosmological role of neutrinos: In May 2009, the ESA satellite PLANCK was launched; combined with astrophysical observations, it promises a significant improvement of the cosmological neutrino mass bounds. The data released in March 2013 report a stringent limit on the sum of neutrino masses of 0.23 eV, which, however, is subject to uncertainties related to the many parameters affecting cosmology. This will, at the same time, improve our knowledge about the role neutrinos played in cosmic evolution. And if neutrinos, via their SUSY partners, the sneutrinos, really are related to the inflationary epoch and acted as key players in the very creation of the universe, PLANCK may help to substantiate these models as well; this knowledge naturally will produce feedback to aid in the comprehension of neutrino properties themselves.

In contrast, the direct detection of the primordial cosmic neutrino background, for example with Tom Weiler's Z-bursts, still seems to lie in the distant future because of insufficient neutrino fluxes of highly energetic cosmic-ray neutrinos. On the other hand, there exists an interesting idea by John F. Beacom and Mark R. Vagins on observing the diffuse background of antineutrinos emitted by supernova explosions. The proposal suggests adding gadolinium in the Super-Kamiokande water tank, which could capture neutrons produced in the inverse beta decay triggered by these neutrinos, and is called GADZOOKS.

Finally, in astrophysics, neutrinos are mutating more and more from research objects to be investigated themselves into probes delivering information about astrophysical processes. The advantage of neutrinos compared with light consists particularly in the property of their being hardly ever absorbed or deflected thanks to their weak interactions. That means that neutrinos can directly image not only processes on the surface, but also processes deep in the interior of stars and supernova explosions. In this sense, neutrinos are like better X-rays. And—a further advantage—the flight direction of neutrinos is not altered by magnetic fields in, for example, the Milky Way, so that neutrinos deliver a faithful instead of a blurred image of astronomical objects.

This research field, *neutrino astrophysics,* started with the measurement of solar neutrinos and the spectacular detection of nineteen neutrino events from a supernova explosion on February 23, 1987, in the detectors IMB and Super-Kamiokande—about three hours before the visible light of the explosion hit the earth. (Unlike light, the neutrinos interact only weakly with dense matter when propagating out from the supernova core. This makes the supernova in its early stages transparent for neutrinos while being opaque for light. The neutrinos thus can escape unhampered while light can escape only when the explosion reaches the collapsing star's surface.) John Learned was one of the members of the IMB team and is, beyond any doubt, one of the brightest and most creative protagonists in neutrino history. Learned, a bandy-legged character with a big bushy beard, slouch hat, and colossal spectacles, who cruises the streets of Honolulu in an ancient Cadillac known as "the whale," was involved in the development of numerous key experiments in neutrino physics. He had started out by sinking strings of photo multipliers—sensitive light detectors—off the Hawaiian shoreline to measure the traces of astrophysical neutrinos. Later Learned proposed, together with Francis Halzen, burying detectors also in the Arctic ice—in holes that he originally intended to fill with alcohol to keep them ice-free so that the detectors could be accessible at any time. The US National Science Foundation (NSF) asked Learned, however, to choose one of the two experiments, and he chose Hawai'i. When Learned lost his first large detector string in the depth of the Pacific Ocean, officials at the NSF may have been relieved that they managed to keep Learned away from the North Pole.

In the meantime, the IceCube Neutrino Observatory is close to full completion in the Antarctic. Almost eighty strings equipped with photo multipliers have been sunk between 1,450 and 2,450 meters deep in the ice around the Amundsen-Scott South Pole station. For comparison, the Eiffel tower is barely 325 meters high. The physicists at the station defy extreme cold (temperatures of minus 40 to minus 80 degrees Celsius), dryness, and, due to an altitude of 3,000 meters, also thin air and disturbed sleep in military tents occupied by twenty people.[1] The only distractions are events like the annual "three-times-around-the-world" run on Christmas day, occasional excursions to a crashed airplane at the end of the runway, and the breathtaking aurora australis in the endless winter night. When leaving the station, they have to stay away from a dangerous area dubbed the "death sector," where the ice could break. The IceCube Neutrino Observatory now constitutes the largest neutrino telescope in the world and has—together with the subsea telescope ANTARES in the Mediterranean surveying the southern hemisphere, and the competing projects NESTOR (Greece) and NEMO (Italy)—a good chance to propel neutrino astrophysics into new realms. As of this writing, in April 2013, the IceCube Collaboration has reported two candidate events, nicknamed Erni and Bert, with PeV (that is, 1,000 TeV) energies that most probably are of extragalactic origin and could be the first indication of an astrophysical neutrino flux. Moreover, extended versions of the IceCube and the KM3NET telescopes (the latter being a joint endeavor of the ANTARES, NEMO, and NESTOR collaborations) with additional detector strings called PINGU and ORCA also have good prospects of solving the issue of neutrino mass hierarchy.

One of the greatest hopes for neutrino astrophysics is to observe neutrinos emitted from a supernova within the Milky Way (Fig. 16.2). Such an event could make possible revolutionary insights in both neutrino physics and astrophysics. From observations in other galaxies one concludes that such a spectacular death of a star should occur in our own galaxy roughly twice a century. The last observation of a supernova in the Milky Way dates back about 400 years to the German astronomer Johannes Kepler. Our own galactic supernova is more than overdue!

The unique ability of neutrinos to penetrate matter has even inspired ideas for useful applications, for instance in communication and imaging as well as in distinct sciences such as geology. For example, the neutrino detectors KamLAND and BOREXINO have recently detected neutrinos created in nuclear decay processes in the earth's interior. Such

Figure 16.2. A supernova remnant in the crab nebula: beacon of hope for neutrino physics and astrophysics. (Courtesy NASA/Defense Video & Imagery Distribution System)

processes contribute to the generation of heat in the deeper layers of earth, which are responsible for both hot lava streams and geothermal energy. A three-dimensional image of the earth's chemical composition might be possible, which would also shed light on the origin and stability of the earth's magnetic field. As early as 1983, the idea of using neutrinos to search for underground sources of oil and ore was advanced by Alvaro De Rujula, Robert Rathbun Wilson (Fermilab founder and nuclear weapon pioneer), and the Nobel Prize winners Georges Charpak and Sheldon Glashow.

The idea of using neutrinos for communication has been discussed in various forms: John Learned, Sandip Pakvasa, and Tony Zee proposed using neutrino beams to send messages to other stellar systems in the Milky Way as well as to search for alien civilizations that might have harnessed neutrinos for communication. Patrick Huber, on the other hand, has imagined utilizing neutrino communication on earth. Modern submarines can remain submerged for very long periods but then have very limited ability to communicate. They return to the sea surface now and then in order to communicate at higher band width. If neutrinos could be used for communication, submarines could remain under water longer. Prasanta Panigrahi and Utpal Sarkar have pursued this idea in reverse, so to speak. They proposed using the neutrino radiation that always accompanies nuclear reactions and can't be shielded in order to spot secret nuclear weapons tests, nuclear submarines, aircraft carriers, or even UFOs. Even more intense neutrino beams could also be used, according to Hirotaka Sugawara, Hiroyuki Hagura, and Toshiya Sanami, for the destruction of nuclear weapons. They argue that neutrinos could be fired at buried armories in order to trigger a hadronic shower in the rock beneath the weapon, which would then induce nuclear fission inside the weapon's material with an explosive power of a few percent of what the weapon was designed for.

Such ideas (some of them admittedly "far out") suggest that neutrinos could develop from objects of basic research into crucial elements of future technologies. But useful application is not the point of neutrino physics. Rather, it is a quest for a better basic understanding of the universe, thereby ultimately contributing to a solid foundation for an incorruptible, rational world view. At times when intolerance, religious fanaticism, gut instincts, and irrational esotericism are flourishing, this is an effort whose importance should not be underestimated. Perhaps in a few years we will know how neutrinos acquire their masses. We may know their exact mass and the origin of the mixing pattern. We may understand what symmetries are responsible for the unique properties of neutrinos. We might even find extra dimensions that could make it possible to send particles back in time. If such a thing turns out to be possible, the neutrino will definitely be the prime candidate to make the journey.

The times remain exciting, and there is a good chance that we may soon discover new physics within the particle desert, a step forward that will inspire us to echo the words of William Clark who, together with Meriwether Lewis, led the first expedition across the North American continent to the Pacific coast: "Ocean in view! O! the joy."

The fundamental symmetries and laws that underlie a theory for all events happening in the universe may reveal themselves a little more completely, the picture behind it all may become a little clearer, and, as sure as death and taxes, neutrinos will contribute in one way or another.

17

Epilogue
Major Tom and the Singing Socrates

When I was ten years old none of the boys in my class liked to dance. The only song that really drove all of us onto the dance floor was "Major Tom" by Peter Schilling. Somehow every one of us could identify with the sad story of the lonely astronaut, whose spaceship suddenly left its intended trajectory and—"drifting, falling, floating, weightless" (in the English version) or "totally detached" (as in the German original)—just like a modern Icarus, followed a mysterious light into the deathly realms of endless space. The last lines are "Now the light commands. This is my home. I'm coming home."[1]

The history of the song is somewhat mysterious; it is brimming with innuendos of altered states of consciousness under the influence of drugs. The character of Major Tom may be based on the US astronaut Major Ed White, who was reluctant to return to his spaceship when he took the first spacewalk in history, or on a Russian story about a young boy who is in love with the moon and thus eventually becomes the cosmonaut Major Tom. As early as 1969, David Bowie wrote and performed the song "Space Oddity" about the astronaut Major Tom. A sequel of the story, in Bowie's song "Ashes to Ashes," contains the lines "Ashes to ashes, funk to funky, we know Major Tom's a junkie."[2]

But whatever booster Major Tom used to shoot himself into his errant trajectory, whatever wave he surfed, on good days a particle physicist can think of himself as being at least a little like Major Tom. Even if a theoretical physicist spends most of his time in an office chair at his desk, his mind is wandering off to the boundaries of space and time, through multiple dimensions of cold emptiness to the hostile heat in the interior of the stars, always on the lookout for something that could be the next hint of the truth lying behind the phenomena. Like the Beat poet Jack Kerouac on his wild car chases across America, physicists on their mind trips also carry "scribbled secret notebooks, and wild typewritten pages, for your own joy," and are "submissive to everything, open, listening."[3]

Not for nothing did Christine Sutton call her own book about the history of neutrino physics *Spaceship Neutrino,* not knowing to which remote places in our wonderful universe this little particle would continue to carry us. Neutrinos are, indeed, as we have seen, the *perfect wave* for such a trip. For the last thirty years there has not been any better one.

And then the other extreme, Socrates, the prototype of a scientist: Plato's teacher, the logic heavyweight among the philosophers of Athens and the mastermind of pure reason.

Physics happens somewhere between these poles. Logical deduction and cold reasoning—even after a night of carousing when everyone else has collapsed below the tables, when the physicist, like Socrates, remains stone-cold sober, without the slightest hint of impairment—or chasing will-o'-the-wisps, like a little kid enamored of the moon, lunatic in its most literal meaning, running after dreams, like Major Tom.

And where else do you find such a broad spectrum of characters? Swashbuckling alpha males like Carlo Rubbia or Hans Volker Klapdor-Kleingrothaus. Brilliant self-promoters like Stephen Hawking or Nino Zichichi. Charismatic teachers like Enrico Fermi. Withdrawn social misfits, so shy that they barely mumble a word, such as Ettore Majorana. Dreamy freaks like Lincoln Wolfenstein. Fighters like Ray Davis. Weirdos and mousy persons in shorts and sandals or short sweaters with coffee stains and ill-fitting trousers that may have come from a recycling bank. Boozing night owls like Wolfgang Pauli and early-bird teetotalers. Full of spirits or overcooled and quiet. Nature lovers like Werner Heisenberg. Cheerful fellows like Einstein. Ascetics. Inventive whiz kids like Cowan and Reines. Relentless calculators like John Bahcall. Scintillatingly witty, creative minds like Nima Arkani-Hamed. Down-to-earth skeptics like—once again—Pauli. Logic destroyers like Kurt Gödel. It seems almost impossible that these people should get along. That things could be peaceful, quite often even cordial. That they can talk to and make sense of each other. But what really counts in physics is neither the temper nor the appearance; above all, it is physics itself.

But what exactly is science? Or more generally: What, after all, is life? "All life is problem solving," according to the British philosopher Karl Popper,[4] and at least on some level, he is obviously right. The problems one solves may arise in daily life, when you decide what to wear, whom to call, where to go, what to buy; or they may show up when dealing with the on-board computer of a spaceship, with a complex experiment in the lab, or when performing tricky calculations on a sheet of paper. Turned around, this means that all life is essentially science: the recognition, analysis, and solving of one problem after another. Popper's philosophy of science was a moderate realism. He believed that the most important task of a scientist was to try to disprove the currently established and generally accepted theories. With every new theory, then, science would become better by a little bit, would get a little closer to the true reality. Later, US philosopher/historian Thomas S. Kuhn started to second-guess this opinion, when he studied real historic processes of scientific revolutions.[5] According to Kuhn's observations, scientists don't try at all to *disprove* generally accepted theories. Rather they *follow* an accepted theory—understood as an exemplary *paradigm*—as long as possible, until so many problems with the theory show up that a crisis is unavoidable. According to Kuhn, it takes this crisis to bring forward the evolution of alternative theories. At a certain point, and not solely driven by rational arguments, the scientists then change the paradigm and jump over to the new theory, to which they now adhere in a new period of normal science until the next crisis leads to another scientific revolution. Kuhn's ideas remind one of the philosophical concept called contructivism,[6] which compares our models of reality with the way a sailor finds his way around some dangerous reefs in the sea. As long as nothing bad happens and the ship doesn't hit a reef and sink, the sailor sticks to his course; he doesn't change his image or model of the seascape, although of course there are other possible ways around the reefs and obstacles. In the extreme case, a scientific model stops claiming any universal truth. Its usefulness is limited to its ability to function like a tool—to make successful predictions, not to reveal general and uniquely true facts about reality.

A common feature of the different opinions about how science works, or should work, is that most important progress in science is a consequence of failure. (Even if Kuhn didn't consider the paradigm change itself as progress, the discovery of obstinate problems in established theory is obviously the most crucial contributor to progress in normal science.) Failure is a fundamental experience of life, and it has been glorified in pop culture, by existentialists and philosophers of life, from Walt Disney's notorious schlemiel, Donald Duck, to Hemingway's old fisherman, who, after eighty-four days of trying, finally catches a fish, the biggest in his entire life, which, after a long fight but before he can land it, he has to witness being ripped apart by hungry sharks.[7] The French author and philosopher Albert Camus declared the Greek hero Sisyphus, the man who was damned in perpetuity to roll a rock up a hill in the netherworld only to see it tumble down again just before he reached the summit, as a role model: "One must imagine Sisyphus happy."[8] And Friedrich Nietzsche eventually understood his *Übermensch,* or superior being, as someone who would defy the absurdity of being and who would accept the aesthetic of the tragedy of his very own failure as the true meaning of life.

In stark contrast to his life-affirming tragic heroes, Nietzsche viewed Socrates as the gravedigger of the ancient world and, at the same time, the prime father of modern science. According to Nietzsche, Socrates, with his rational questioning, anticipated the scientific worldview and destroyed the immediate Dionysian unity with the world, a "despotic logician who never glowed with the artist's divine frenzy."[9] For Nietzsche, science thus was the original cause for the ancient Greeks' suffering from the separation of subject and object. And for modern humanity being torn from the integral unity of nature.

Only if science were reconciled with the Dionysian tragedy, with art and music, only if Socrates would start to *make music,* only then could science grant humanity with a deep metaphysical benefit, could establish a true meaning of life.

In view of this, it is probably most amazing that science itself, with the many-worlds interpretation of quantum mechanics, leaves some room for an entire multiverse of alternative realities beyond our immediate experience—a multiverse mirroring the unity of Parmenides's philosophy, as well as the egolessness of Aldous Huxley under the influence of psychoactive drugs, and a Dionysian creative moment bursting open all boundaries. And even if it is not clear whether intratheoretical concepts such as the multiverse of quantum physics can be compared in a meaningful way with meta-theoretic concepts such as alternative theories within the philosophy of constructivism, maybe it is this very property—that the theory is open, that it defines its own limits—that prompts the scientific worldview to reach beyond its significance as a mere tool, and lets it feel a scent of the true reality, much in the way that a masterpiece of art—as revealed by both Vincent van

Figure 17.1. Van Gogh's *The Starry Night:* more true than reality itself.

Gogh[10] and Ernest Hemingway[11]—can be *more true than reality itself* (see Fig. 17.1).

And last but not least: Physical laws are more than a mere bunch of unintuitive equations and parameters. They are images of nature, and, just like the masterpieces of art, the theories of modern physics astonish with their enchanting aesthetic when fundamental laws of nature are traced back to symmetries, Platonic shapes, and warped geometries, such as the crazy buildings of the Spanish shooting-star architect Santiago Calatrava. Images that can be approached more closely with the help of the "Keys to the Universe,"[12] the neutrinos, which now appear to us a little less blurry after the revolutionary developments in neutrino physics in recent years: Science is Art!

And eventually, according to Nietzsche and Popper: Art makes life meaningful—Life is Science—Science is Art—SOCRATES IS SINGING . . .

Notes

1. Dawn Patrol in Honolulu

1. Steven Kotler, *West of Jesus* (New York: Bloomsbury, 2006).

2. Eleusis, Plato, Magic Mushrooms

1. Friedrich Nietzsche, *The Birth of Tragedy: Out of the Spirit of Music* (New York: Penguin Classics, 1994).
2. *Homeric Hymn to Demeter,* trans. from the Greek by Hugh G. Evelyn-White and first published by the Loeb Classical Library in 1914, earlywomenmasters.net/demeter/myth_470.html.
3. Albert Hofmann, *LSD, My Problem Child: Reflections on Sacred Drugs, Mysticism, and Science* (Oxford: Oxford University Press, 2013); Marion Giebel, *Das Geheimnis der Mysterien* (Mannheim: Artemis, 2003); Carl Kerenyi, *Eleusis: Archetypical Image of Mother and Daughter* (Princeton: Princeton University Press, 1991).
4. Carl A. P. Ruck, in R. Gordon Wasson, Albert Hofmann, and Carl A. P. Ruck, *The Road to Eleusis* (New York: Harcourt Brace Jovanovich, 1978).
5. Kerenyi, *Eleusis.*
6. Hofmann, *LSD.*
7. William Blake, *The Marriage of Heaven and Hell* (Oxford: Bodleian Library, University of Oxford, 2011).
8. The Doors, "Break on Through" (Hollywood: Sunset Sound Recorders, 1966).
9. Aldous Huxley, *The Doors of Perception: Heaven and Hell* (New York: Harper Colophon Books, 1963).
10. Hofmann, *LSD.*
11. Wasson, Hofmann, and Ruck, *Road to Eleusis.*
12. Nietzsche, *Birth of Tragedy.*
13. Hofmann, *LSD.*
14. Giebel, *Mysterien.*

15. Jürgen Aschaff et al., *Die Zeit—Dauer und Augenblick* (Munich: Piper, 1989).

3. Quantum Physics

1. Werner Heisenberg, *Physics and Beyond: Encounters and Conversations* (New York: Harper & Row, 1972).
2. Armin Hermann, *Rowohlt Monographie Werner Heisenberg* (German edition) (Reinbek: Rowohlt, 2001).
3. Thomas Powers, *Heisenberg's War: The Secret History of the German Bomb* (New York: Knopf, 1993).
4. Paul Lawrence Rose, *Heisenberg and the Nazi Atomic Bomb Project, 1939–1945: A Study in German Culture* (Berkeley: University of California Press, 2001); Jeremy Bernstein, *Hitler's Uranium Club: The Secret Recordings at Farm Hall* (New York: Springer, 2000).
5. Powers, *Heisenberg's War*.
6. "Comment by Jochen Heisenberg," wernerheisenberg.unh.edu /washsys.htm.
7. Michael Frayn, *Copenhagen* (New York: Anchor, 2000).
8. Hermann, *Rowohlt*.
9. Richard Feynman, *The Character of Physical Law* (Cambridge, MA: MIT Press, 1967).
10. Werner Heisenberg, *Der Teil und das Ganze* (Munich: Piper, 2002), my translation.
11. Erwin Schrödinger, "Die gegenwärtige Situation in der Quantenmechanik" [The present situation in quantum mechanics], *Naturwissenschaften* 23, no. 48 (1935): 807–812, trans. John D. Trimmer, in *Proceedings of the American Philosophical Society* 124 (1980): 323–338.
12. R. Corby Hovis and Helge Kragh, "P.A.M. Dirac and the Beauty of Physics," *Scientific American* 268, no. 5 (May 1993): 104–109.
13. Thomas S. Kuhn, *The Structure of Scientific Revolutions* (Chicago: University of Chicago Press, 2012).
14. Hermann, *Rowohlt*.
15. Erwin Schrödinger, *Nature and the Greeks* and *Science and Humanism* (Cambridge: Cambridge University Press, 1996).
16. Hovis and Kragh, "P.A.M. Dirac."

17. Johann W. von Goethe, "Natur und Kunst," trans. Robert J. Richards, in Robert J. Richards, "Nature Is the Poetry of Mind, or How Schelling Solved Goethe's Kantian Problems," home. uchicago.edu/~rjr6/articles/Schelling-Goethe.pdf.

18. Carl Friedrich von Weizsäcker, *The Unity of Nature* (New York: Farrar, Straus & Giroux, 1981).

19. A. H. Coxon, *The Fragments of Parmenides: A Critical Text with Introduction and Translation, the Ancient Testimonia and a Commentary* (Las Vegas: Parmenides Publishing, 2009).

20. Parmenides, "On Nature," philoctetes.free.fr/parmenides.pdf.

21. Huxley, *Doors of Perception.*

22. Carl Friedrich von Weizsäcker, "Parmenides und die Quantentheorie," in *Die Einheit der Natur* (Munich: Hanser-Verlag, 1982).

23. Plato, *Parmenides* (Kittening: Arc Manor, 2008).

24. Luciano De Crescenzo, *Geschichte der griechischen Philosophie* (Zurich: Diogenes, 1985), my translation.

25. Weizsäcker, "Parmenides."

26. Ibid.

27. Ibid., my translation.

28. William Hermanns, *Einstein and the Poet: In Search of the Cosmic Man* (Wellesley: Branden Books, 1983).

29. Heraclitus, "Fragment 52," cited in the Wikipedia article "Heraclitus," en.wikipedia.org/wiki/heraclitus#cite_note-53.

30. Peter Aldous, "Interview with Mark Oliver Everett (E)," *New Scientist,* November 24, 2007.

31. Douglas Adams, *The Hitchhiker's Guide to the Galaxy* (New York: Harmony Books, 1979).

4. Black Dots on a White Background

1. Betrand Russell, *A History of Western Philosophy* (New York: Simon & Schuster / Touchstone, 1967).

2. Betrand Russell, *Denker des Abendlandes* (Bindlach: Gondrom, 2000), my translation.

3. Gary Taubes, *Nobel Dreams: Power, Deceit, and the Ultimate Experiment* (New York: Random House, 1987).

4. Ibid.

5. Beyond the Desert

1. King James Bible, Exodus 3:8.
2. Auguste Dick, *Emmy Noether, 1882–1935 (Beihefte zur Zeitschrift Elemente der Mathematik)* (Basel: Birkhäuser, 1970).
3. Plato, *Timaios* (German edition) (Ditzingen: Reclam 2003), my translation.
4. Werner Heisenberg, Hans-Peter Dürr, Walter Blum, and Helmut Rechenberg, *Werner Heisenberg, Gesammelte Werke. Collected Works*, part C, vol. 3: *Physik und Erkenntnis 1969–1976* (Munich: Piper, 1985).
5. Adams, *Hitchhiker's Guide*.
6. Hans-Volker Klapdor-Kleingrothaus and Heinrich Päs, *Beyond the Desert: Workshop on Particle Physics beyond the Standard Model* (Bristol, UK: IOP Publishing, 1998).

6. From Symmetry Breaking to Supersymmetry

1. "Henry Cowell Piano Music," *Three Irish Legends,* cowellpiano .com/TIL.html.
2. Rolf Heuer, press conference and interview, Deutsche Welle (Germany), www.dw.de/cern-director-general-says-the-work -starts-now/a-16072369.
3. Plato, *The Republic* (Hollywood, FL: Simon & Brown, 2013).
4. Abdus Salam, "Gauge Unification of Fundamental Forces," Nobel Lecture, December 8, 1979, www.nobelprize.org/nobel_prizes /physics/laureates/1979/salam-lecture.html.
5. Ibid.
6. Ibid.
7. Brian Greene, *The Elegant Universe: Superstrings, Hidden Dimensions, and the Quest for the Ultimate Theory* (New York: Vintage Books, 2000).

7. Birth of an Outlaw

1. Christine Sutton, *Spaceship Neutrino* (Cambridge: Cambridge University Press, 1992).
2. I owe this comparison to Yorck Ramachers, a member of our Heidelberg research group and now professor at the University of Warwick, UK.

3. David Lindorf, *Pauli and Jung: The Meeting of Two Great Minds* (Wheaton, IL: Quest Books, 2004).

4. Sutton, *Spaceship Neutrino*.

5. Jan Philipp Bornebusch, "Das 'Gewissen' der Physik—Am 15. Dezember 1958 starb Wolfgang Pauli," *Spektrum Direkt*, www .spektrum.de/alias/quantenphysik/das-gewissen-der-physik /976922.

6. Fred Hoyle, Bibliography, *Proceedings of the Royal Society*, vol. A, 301 (1967): 171.

7. Necia Grant Cooper, ed., "Celebrating the Neutrino," *Los Alamos Science* 25 (1997).

8. Ibid.

9. Leonardo Sciascia, *Das Verschwinden des Ettore Majorana* (Berlin: Wagenbach, 2003).

10. Ibid.

11. Ibid.

12. Tommaso Dorigo, "Ettore Majorana: The Mystery Might Be Solved," www.science20.com/quantum_diaries_survivor/ettore_majorana _mystery_might_be_solved-79823.

13. Oleg B. Zaslavskii, "Quantum Mechanics of Destiny," *Priroda* 11 (2006): 5–63; e-print arxiv.org/abs/physics/0605001.

14. E. Akhmedov, private communication.

9. New Physics Is Falling from the Skies

1. Takaaki Kajita, talk at the Eighteenth International Conference on Neutrino Physics and Astrophysics (NEUTRINO '98), Takayama, Japan, June 4–9, 1998.

2. Hitoshi Murayama, "The Origin of Neutrino Mass," *Physics World* (May 2002): 35.

3. U.S. President Bill Clinton at the MIT commencement, June 6, 1998, clinton4.nara.gov/WH/New/html/19980605-28045. html.

4. Richard P. Feynman, *What Do You Care What Other People Think?* (New York: W. W. Norton, 2001).

5. Ray Davis, "A Half-Century with Solar Neutrinos," Nobel Lecture, December 8, 2002, www.nobelprize.org/nobel_prizes/physics/laureates /2002/davis-lecture.pdf.

6. Wick Haxton, "The Scientific Life of John Bahcall," *Annual Review of Nuclear and Particle Science* 59 (2009): 1–20, e-print arxiv.org/abs /0904.2865.
7. Palash B. Pal, private communication.
8. Alexei Yu. Smirnov, private communication.
9. John Learned, "Discovery of Neutrino Mass and Oscillations," www.phys.hawaii.edu/~jgl/nuosc_story.html.
10. Arthur McDonald, private communication.

10. Cosmic Connections
1. Steven Weinberg, *The First Three Minutes* (New York: Basic Books 1993).
2. www.boston.com/realestate/galleries/springsweep/13.htm.
3. Alan Guth, *The Inflationary Universe* (New York: Basic Books, 1998).
4. Martin Heidegger, *Sein und Zeit* (Tübingen: Niemeyer 2006), trans. Michael Inwood, "Does the Nothing Noth?" in *German Philosophy since Kant*, ed. Antony O'Hear (Cambridge: Cambridge University Press, 1999).

11. Neutrinos
1. Peter Minkowski, private communication.

12. Extra Dimensions, Strings, and Branes
1. ScienceWatch interview with Nima Arkani-Hamed, www.esi -topics.com/brane/interviews/DrArkani-Hamed.html.
2. Nima Arkani-Hamed, Savas Dimopoulos, and Georgi Dvali, "The Universe's Unseen Dimensions," *Scientific American* 283, no. 2 (August 2000): 62–69.
3. Daniela Wünsch, "Der Erfinder der fünften Dimension," *Neue Züricher Zeitung*, January 21, 2004, termessos.de/PIX/Kaluza/NZZ _Onlineed.html, my translation.
4. Greene, *Elegant Universe*.
5. Peter Woit, *Not Even Wrong: The Failure of String Theory and the Search for Unity in Physical Law* (New York: Basic Books, 2007).
6. Leonard Susskind, *The Cosmic Landscape: String Theory and the Illusion of Intelligent Design* (Boston: Back Bay Books, 2006).
7. Goran Senjanovic, private communication.

8. Lisa Randall, *Warped Passages: Unraveling the Mysteries of the Universe's Hidden Dimensions* (New York: HarperPerennial, 2006).

13. Einstein's Heritage

1. Armin Herrmann, *Einstein—der Weltweise und sein Jahrhundert* (Munich: Piper, 1994).
2. Leonard Susskind and James Lindesay, *An Introduction to Black Holes, Information, and the String Theory Revolution: The Holographic Universe* (Singapore: World Scientific, 2005).
3. It is often questioned whether Einstein indeed knew of the Michelson-Morley-experiment, a question addressed by Jeroen van Dongen, "On the Role of the Michelson-Morley Experiment: Einstein in Chicago," e-print arxiv.org/abs/0908.1545.
4. Compare Matt Visser, *Lorentzian Wormholes: From Einstein to Hawking* (College Park, MD: American Institute of Physics, 1996).

14. How to Build a Time Machine

1. Paul Yourgrau, *A World without Time: The Forgotten Legacy of Gödel and Einstein* (New York: Basic Books, 2006).
2. Jim Holt, "Time Bandits: What Were Einstein and Gödel Talking About?" *New Yorker,* February 28, 2005.
3. Frank J. Tipler, "Rotating Cylinders and the Possibility of Global Causality Violation," *Physical Review D* 9 (1974): 2203–2206.
4. Kip Thorne, *Black Holes and Time Warps: Einstein's Outrageous Legacy* (New York: W. W. Norton, 1995).
5. Robert A. Heinlein, *The Fantasies of Robert A. Heinlein* (New York: Tor Books, 2002).
6. "Avengers Forever Villains," marvelite.prohosting.com/surfer/cmarvel /afvillains.html.
7. Stephen W. Hawking, "The Chronology Protection Conjecture," *Physical Review D* 46 (1992): 603–611.
8. Stephen W. Hawking, "Space and Time Warps," lecture, hawking .org.uk/space-and-time-warps.html.

15. Against Hawking and the Timekeepers

1. Dan Brown, *Angels and Demons* (New York: Atria, 2003).
2. Visser, *Lorentzian Wormholes.*
3. Leonard Susskind, e-print arxiv.org/abs/gr-qc/0504039.

4. Visser, *Lorentzian Wormholes.*
5. Marcus Chown, "The Extra-dimensional Time Machine," *BBC Focus,* October 2006, 52–56.
6. Joe Haldeman, *The Accidental Time Machine* (New York: Ace Hardcover, 2007).
7. Mark Alpert, *Final Theory* (New York: Touchstone, 2008).
8. David Deutsch and Michael Lockwood, "The Quantum Physics of Time Travel," *Scientific American* 270, no. 3 (1994): 68–74.

16. Into the Wilderness of the Terascale

1. Interview with Torsten Schmidt, www.desy.de/expo2000/deutsch /dhtmlbrowser/webthemen/21_amanda/antarctica.htm 126.

17. Epilogue

1. Peter Schilling, "Major Tom (Coming Home)" (New York: Elektra Records, 1983).
2. David Bowie, "Ashes to Ashes" (New York: RCA Records, 1980).
3. Jack Kerouac, "Belief and Technique for Modern Prose," 1959, www.writing.upenn.edu/~afilreis/88/kerouac-technique.html.
4. Karl Popper, *All Life Is Problem Solving* (London: Routledge, 2001).
5. Thomas S. Kuhn, *The Structure of Scientific Revolutions* (Chicago: University of Chicago Press, 1962).
6. Paul Watzlawik, ed., *Die erfundene Wirklichkeit* (Munich: Piper, 1981).
7. Ernest Hemingway, *The Old Man and the Sea* (New York: Scribner, 1996).
8. Albert Camus, *The Myth of Sisyphus* (New York: Penguin, 2005).
9. Nietzsche, *Birth of Tragedy.*
10. Ulrich M. Schneede, *Vincent van Gogh: Leben und Werk* (Munich: C. H. Beck, 2003).
11. Ernest Hemingway, "Old Newsman Writes," *Esquire,* December 1934, 26.
12. Cover, *New Scientist* 2615, August 4, 2007.

Further Reading

Neutrinos

Christine Sutton, *Spaceship Neutrino* (Cambridge: Cambridge University Press, 1992).

A comprehensive account of neutrino physics up to the early 1990s, as seen by a well-known experimentalist. It is an excellent source of information, although one is sometimes overwhelmed with technical details. Not covered are the exciting developments in neutrino oscillations that started in the mid-1990s.

Necia Grant Cooper, ed., "Celebrating the Neutrino," *Los Alamos Science* 25 (1997).

A collection of articles by scientists who were involved, covering some of the moments of glory in neutrino physics.

Frank Close, *Neutrino* (Oxford: Oxford University Press, 2010).

A modern book about neutrinos by the well-known particle physicist and science writer Frank Close, with only limited attention to the neutrino's connection to new physics.

Leonardo Sciascia, *The Moro Affair and the Mystery of Majorana* (New York: HarperCollins, 1991).

A novel exploring the puzzling disappearance of the brilliant physicist Ettore Majorana. It was a best seller in Italy and contributed more to the fame of its protagonist in his home country than his groundbreaking works on quantum, nuclear, and neutrino physics.

Joao Magueijo, *A Brilliant Darkness: The Extraordinary Life and Mysterious Disappearance of Ettore Majorana, the Troubled Genius of the Nuclear Age* (New York: Basic Books, 2009).
    The author of this book about Majorana is a well-known cosmologist.

Quantum Theory and Werner Heisenberg

Werner Heisenberg, *Physics and Beyond: Encounters and Conversations* (New York: Harper & Row, 1972).
    Heisenberg writes about the evolution of physics in a manner not unlike the way Plato writes about philosophy. He reviews conversations with Einstein, Bohr, and Dirac, as well as with boyhood friends and political opponents— about science, religion, philosophy, and, again and again, about quantum mechanics. He focuses more on his own life and philosophy than on pure science. Despite that, or actually because of it, this is an extremely interesting historical document, and one of the greatest books about physics and how it is pursued.

Armin Hermann, *Rowohlt Monographie Werner Heisenberg* (German edition) (Reinbek: Rowohlt, 2001).

Nice Heisenberg biography with many original quotes demonstrating Heisenberg's enthusiasm for physics, nature, and Platonic philosophy.

David C. Cassidy, *Uncertainty: The Life and Science of Werner Heisenberg* (New York: W. H. Freeman, 1993).
A well-respected biography of Heisenberg.

David C. Cassidy, *Beyond Uncertainty: Heisenberg, Quantum Physics, and the Bomb* (New York: Bellevue Literary Press, 2009).
An update of his earlier biography, after the Farm Hall transcripts were released, focusing more on Heisenberg's role during World War II.

Thomas Powers, *Heisenberg's War: The Secret History of the German Bomb* (New York: Knopf, 1993).
A controversial account that glorifies Heisenberg's role in the German nuclear-energy project during World War II. The conclusions are to be taken with a grain of salt.

Kenneth W. Ford, *The Quantum World: Quantum Physics for Everyone* (Cambridge, MA: Harvard University Press, 2004).
A beautiful and lucid introduction to the quantum world that provides you with a sense of dimensions and concepts and makes you see particles like friends or family members.

Kenneth W. Ford, *101 Quantum Questions: What You Need to Know about the World You Can't See* (Cambridge, MA: Harvard University Press, 2012).

A recent sequel that also touches on advanced topics such as superconductivity, Hawking radiation, and the Higgs boson.

Marcus Chown, *The Quantum Zoo: A Tourist's Guide to the Never-Ending Universe* (Washington, DC: Joseph Henry Press, 2006);
Marcus Chown, *The Universe Next Door: The Making of Tomorrow's Science* (Oxford: Oxford University Press, 2003).
Chown writes clearly and entertainingly about paradoxes in quantum mechanics and other popular science.

Particle Physics

A. Zee, *Fearful Symmetry* (New York: Macmillan, 1986).
Tony Zee is not only one of the most creative particle physicists, he is also a true renaissance man and writes with remarkable wit. This book about beauty in physics is definitely worth reading.

Dan Hooper, *Nature's Blueprint: Supersymmetry and the Search for a Unified Theory of Matter and Force* (Washington, DC: Smithsonian, 2008).
Hooper's book about supersymmetry describes the most popular extension of the Standard Model, which is relevant for searches at the LHC.

Don Lincoln, *The Quantum Frontier: The Large Hadron Collider* (Baltimore: Johns Hopkins University Press,

2009); and Gian Francesco Giudice, *A Zeptospace Odyssey: A Journey into the Physics of the LHC* (Oxford: Oxford University Press, 2010).

Two recent books about the LHC, the most important particle physics endeavor of all time.

Sean Carroll, *The Particle at the End of the Universe: How the Hunt for the Higgs Boson Leads Us to the Edge of a New World* (New York: Dutton Adult, 2012).

One of the first books about the Higgs to appear after its probable discovery in 2012.

Relativity and Time Travel

Kip Thorne, *Black Holes and Time Warps: Einstein's Outrageous Legacy* (New York: W. W. Norton, 1995).

Thorne has pursued John Wheeler's wormholes further than anyone and is a pioneer in studying gravity waves. He is one of the most important researchers into the physics of gravity. Generations of physicists also have learned general relativity from the classic textbook he wrote with Charles Misner and John Wheeler (the cover decorated with the apple that supposedly once fell on Newton's head, inspiring him to think about gravity). Thorne's popular science book describes the history of relativity with several anecdotes and personal experiences, written by one who really knows it.

J. Richard Gott, *Time Travel in Einstein's Universe: The Physical Possibilities of Travel through Time* (Boston: Mariner Books, 2002).

A popular science book specifically about the possibilities of time travel, written by one of the protagonists of the field, the Princeton cosmologist Richard Gott, who believes that the entire universe may have been created in a time loop.

Armin Herrmann, *Einstein—der Weltweise und sein Jahrhundert* (Munich: Piper, 1994).
The German Einstein biography I mainly used as a source.

Abraham Pais, *Subtle Is the Lord: The Science and the Life of Albert Einstein* (Oxford: Oxford University Press, 2005).
A classic Einstein biography.

Paul J. Nahin, *Time Machines: Time Travel in Physics, Metaphysics, and Science Fiction* (New York: Springer, 1998).
Everything about time travel: movies, science fiction novels, cartoons, and physics with formulas and diagrams. A great source of information on the topic.

Sean Carroll, *From Eternity to Here: The Quest for the Ultimate Theory of Time* (New York: Plume, 2010).
Beautiful book about the concept of time in physics.

Cosmology

Alan Guth, *The Inflationary Universe* (New York: Basic Books, 1998).

This book is to cosmology what Thorne's book is to general relativity: a comprehensive account and an entertaining read, full of anecdotes and memories by the father of inflation himself.

Steven Weinberg, *The First Three Minutes* (New York: Basic Books, 1993).
Nobel Prize winner Weinberg is one of the great minds of physics, still an instigator of new ideas. His book is one of the first popular science books about cosmology: full of solid expertise, albeit now somewhat outdated, and sometimes dry.

Dan Hooper, *Dark Cosmos: In Search of Our Universe's Missing Mass and Energy* (New York: HarperPerennial, 2007).
A modern book about cosmology, concentrating on dark matter and dark energy.

Superstrings and Extra Dimensions

Brian Greene, *The Elegant Universe: Superstrings, Hidden Dimensions, and the Quest for the Ultimate Theory* (New York: Vintage Books, 2000).
An extensive account of the most promising approach to a theory of everything. It is amazingly detailed without ever getting blurry or boring. This is the book about superstrings that is comparable to Thorne's about relativity or Guth's about cosmology.

Lisa Randall, *Warped Passages: Unraveling the Mysteries of the Universe's Hidden Dimensions* (New York: HarperPerennial, 2006).
A book specifically about extra dimensions, by one of the major players in the field.

Daniela Wuensch, *Der Erfinder der 5. Dimension. Theodor Kaluza: Leben und Werk* (Göttingen: Termessos, 2009).
The significance of Kaluza's work is just starting to get the attention it deserves. A first biography in German.

Leonard Susskind, *The Cosmic Landscape: String Theory and the Illusion of Intelligent Design* (Boston: Back Bay Books, 2006);
Leonard Susskind, *The Black Hole War: My Battle with Stephen Hawking to Make the World Safe for Quantum Mechanics* (New York: Little, Brown, 2008); and
Leonard Susskind and James Lindesay, *An Introduction to Black Holes, Information, and the String Theory Revolution: The Holographic Universe* (Singapore: World Scientific, 2005).
Two popular science books and a physics textbook about two extremely topical ideas that presently feature in the discussions of the best theoretical physicists in the world. The author as originator of these and many other important concepts in modern physics is one of the most creative scientists of his time, and also an exceptionally gifted narrator. Susskind's two popular books are witty, enthusiastic, and clear, and are, next to Heisenberg's autobiography, two

of the most beautiful books about how physics is actually pursued.

Philosophy of Science

Karl Popper, *The Logic of Scientific Discovery* (London: Routledge, 2002).
A classic of the philosophy of science, famous for its clarity.

Thomas S. Kuhn, *The Structure of Scientific Revolutions* (Chicago: University of Chicago Press, 1962).
An inspiring and influential book that introduced the word "paradigm" into intellectual discourse. Kuhn looks into the history of science in order to understand how science actually works, and finds out that it is less rational than often claimed. Even if you disagree, a great read.

Paul Watzlawik, ed., *Die erfundene Wirklichkeit* (Munich: Piper, 1981).
A collection of papers about the concept of radical constructivism/constructivist epistemology.

Ernst von Glasersfeld, *Radical Constructivism* (London: Routledge, 1996).
A book on constructivist epistemology by one of the major players in the field.

Plato, Nietzsche, and Philosophy

Plato, *Timaeus* (Newburyport, MA: Focus Publishing, 2001).
The Socratic dialogue that is most important for phys-
ics. It introduces the concept of symmetries and influenced
Heisenberg.

Plato, *Parmenides* (Kittening: Arc Manor, 2008).
The Socratic dialogue that discusses the unity of all that
is. Highly influential for all western metaphysics, but difficult to understand and read.

Friedrich Nietzsche, *The Birth of Tragedy: Out of the Spirit of
Music* (New York: Penguin Classics, 1994).
Nietzsche's first book, his hymn to the spirit of ancient
Greece. Written sweepingly, sometimes even bordering on
psychotic, it has been influential in all parts of arts and
culture. One of the favorite books of Jim Morrison, lead
singer of the 1960s rock band the Doors.

Jürgen Aschaff et al., *Die Zeit—Dauer und Augenblick*
(Munich: Piper, 1989).
A collection of articles from different disciplines about
the concept of time. A great source of information.

Bertrand Russell, *A History of Western Philosophy* (New
York: Simon & Schuster / Touchstone, 1967).
Still one of the most beautiful books about the history of
philosophy.

Carl Friedrich von Weizsäcker, *The Unity of Nature* (New York: Farrar, Straus & Giroux, 1981).
A collection of articles by a physicist and philosopher who was among the best in both fields. The intriguing links between the disciplines are worked out clearly. It is well written and never gets dry.

Eleusis and Psychedelic Drugs

Albert Hofmann, *LSD, My Problem Child: Reflections on Sacred Drugs, Mysticism, and Science* (Oxford: Oxford University Press, 2013).
The fantastic adventures of a conservative Swiss chemist whose accidental discovery of LSD changed his life. Reads like a novel.

Marion Giebel, *Das Geheimnis der Mysterien* (Mannheim: Artemis, 2003).
A book about the history of mystery cults in ancient Greece.

Carl Kerenyi, *Eleusis: Archetypical Image of Mother and Daughter* (Princeton: Princeton University Press, 1991).
Interesting classic text about the mystery cult in Eleusis.

Carl A. P. Ruck, in R. Gordon Wasson, Albert Hofmann, and Carl A. P. Ruck, *The Road to Eleusis* (New York: Harcourt Brace Jovanovich, 1978).

The book in which the Swiss chemist and discoverer of LSD, Hofmann, teams up with a classical scholar and a mushroom specialist to unravel the secret of Eleusis. Controversial but enthralling.

Solomon H. Snyder, *Drugs and the Brain* (New York: W. H. Freeman, 1996).
Beautiful book about the effects of psychoactive drugs on the brain, from heroin to cocaine to LSD.

Aldous Huxley, *The Doors of Perception: Heaven and Hell* (New York: Harper Colophon Books, 1963).
Philosophic insights and experiences about psychedelic drugs by a real master.

Fiction

Mark Alpert, *Final Theory* (New York: Touchstone, 2008).
Science thriller by *Scientific American* editor Mark Alpert, about sterile neutrinos in extra dimensions.

Joe Haldeman, *The Accidental Time Machine* (New York: Ace Hardcover, 2007).
Science fiction story about time travel, with a reference to time-travelling neutrinos in the afterword.

Magazine Articles

Mark Alpert, "Dimensional Shortcuts," *Scientific American,* August 2007.

Marcus Chown, "The Extra-dimensional Time Machine," *BBC Focus,* October 2006, 52–56.

Marcus Chown, "Head 'em off at the Past," *New Scientist* 2552, May 20, 2006, 34.

Marcus Chown, "Neutrinos: The Key to a Theory of Everything," *New Scientist* 2615, August 4, 2007.

Martin Hirsch, Heinrich Päs, and Werner Porod, "Ghostly Beacons of New Physics," *Scientific American,* April 2013, 41–47.

John G. Kramer, "Back in Time through Other Dimensions," *Analog Science Fiction & Fact,* October 2006, 86.

Acknowledgments

It is sometimes difficult to convey the fascination a tiny particle, the neutrino, can have on you. I thus would like to express my deep gratitude to the many people who have helped me to write and publish this book.

Michael G. Fisher at Harvard University Press shared my enthusiasm from the first moments on and was a great help in all stages of the production process. Kenneth W. Ford was a fantastic editor who corrected not only my less-than-perfect English but also many physics and history facts. He definitely made this a much better book than it was before, and he leaves me greatly impressed by his expertise and attention to detail. Lauren K. Esdaile patiently worked with me on all technical issues, from file uploads to figure permissions to contracts.

The English edition is based on an earlier German edition, and I once more have to thank the great support I received from my German publisher, Piper, and in particular from Katharina Wulffius, Anne Wiedemeyer, and Wolfgang Gartmann. I also thank my Dortmund colleague Metin Tolan, who first established the contact with Piper.

Earlier drafts of the manuscript have been read by Annette Krais, Arnulf Krais, Monika Nolden, Annika Päs, and Alexa Schieck, all of whom gave me helpful comments. Yasmin

Anstruther, Frank Deppisch, Martin Hirsch, Alexander Lenz, and Bela Majorovits have read the complete manuscript and made me aware of several mistakes and points that were not clear. Bela Majorovits, Susanne Mertens, Stefan Schönert, and Herbert Strecker supplied me with photos and information about the experiments GERDA and KATRIN. Martin Hirsch, David Cline, Luigi di Lella, John Learned, Art McDonald, Sandip Pakvasa, Bernhard Stech, Karl Strobel, and Tom Weiler have provided me with further important facts. Finally, both the German and the English language editions of Wikipedia were an invaluable source of information.

This book would not have been written without my scientific collaboration with Sandip Pakvasa, Tom Weiler, and James Dent on shortcuts in extra dimensions and neutrino time travel.

Finally, Klaus and Silvia Heising and Matthias Nennecke were a great help to my family during the stressful times when I was working on the German edition of this book.

What is true for books also applies to science: Success is almost always the result of a collaborative effort. But anyone who wants to tell an enthralling story necessarily has to concentrate on the fate of particular individuals and his own subjective experience, just to avoid drowning in an ocean of names and facts. So I apologize to anyone who—as a result of my storytelling, my ignorance, or my vanity—finds herself or himself described incorrectly, too briefly, or not at all. In particular I ask for understanding that I have focused on my own contributions, views, and experiences. After all, even in *The*

Lord of the Rings, an epic story is told from the viewpoint of a lowly hobbit.

Above all I hope I have written a book that is fun to read and that whets the reader's appetite for more. If the reader develops her or his own different viewpoint about the story of neutrinos, nobody will be happier than I.

Index

Note: Page numbers in *italics* refer to illustrations. Page numbers followed by the letter *t* refer to tables.

active galactic nuclei, 125
Adams, Douglas, 33, 59
ADD, 177–182, *183,* 184–186
Akhmedov, Evgeny, 165
Akulov, Vladimir, 76
ALICE detector, 47
Alice in Wonderland, 170, 183
"All," 26, 27, 229
Allanach, Benjamin, 235
allegory of the cave, 15, 69–70
Alpert, Mark, 228
Alpher, Ralph, 134
Amundsen-Scott South Pole Station, 243
Anaxagoras, 34
Andromeda Nebula, 132
angular momentum conservation, 36, 56
ANTARES experiment, 244
anthropic principle, 176
antimatter asymmetry, 144
antineutrino
    detection of, *85*
    right-handed, 104
antiparticles, 38–39, 91
antiprotons, 144
Antoniadis, Ignatius, 177, 179
Antusch, Stefan, 166

Arkani-Hamed, Nima, 168–169, 176, 177, 250
    See also ADD
ATLAS detector, 47, *181, 232–233*
atmospheric neutrinos, 126, 216, 218
atomic nucleus. *See* nucleus, atomic
atomism, 15–16, 29, 35
atoms, 35
Aulakh, Charanjit Singh, 162
Aurelius, Marcus, 8

Babu, Kaladij, 112
background radiation. *See* CMB; cosmic neutrino background
background radioactivity, 99, 102–103, 110–111
backward in time. *See* time travel
Baer, Howard, 64
Bahcall, John, 93, 119–121, 250
Barger, Vernon, 122, 130
baryogenesis, 153, 164
baryon asymmetry, 164
baryons, 41
    conservation of, 144
Baudis, Laura, 111
Beacom, John F., 241
beauty in science, 25, 52, 54, 78, 254, 255

beta decay, 42, 57, 82, 92, 104
  asymmetry in, 74
  double, 105, 106, 107, 113, 148,
    238
  Fermi theory of, 83
  inverse, *85,* 241
  and neutrino oscillation, 95
  See also neutrinoless double
    beta decay
Bhattacharyya, Gautam, 162, 163,
  182, 216, 236
Big Bang, 132, 135, 141
  nucleosynthesis, 146
bi-maximal mixing, 130
black hole
  coinage, 30
  microscopic, 180, *181,* 237
Blake, William, 12
Bohr, Niels
  and Everett, 30
  and Heisenberg, 18–19, 23, 25
  in Moscow, 24
  See also Copenhagen
    interpretation
bomb, nuclear, 83, 93, 94
  and Heisenberg, 18
  and neutrino hunt, 84–86
bootstrap paradox, 210, 228
BOREXINO experiment, 244
Born, Max, 83
Bose, Satyendra Nath, 40
bosons, 40
Bowie, David, 248
branes, 173–175, 219, 224
  oscillating, 221
Brimos (Greek god), 11
Brout, Robert, 69
Brown, Dan, 215
Buchmüller, Wilfried, 165
Bukowski, Charles, 34
bulk (non-brane), *183*
Buniy, Roman, 211

Cabibbo, Nicola, 123
Cabrera, Blas, 136
Calatrava, Santiago, 254
Camus, Albert, 252
Casas, J. Alberto, 155
Cauet, Christophe, 234
causality, breakdown of, 21, 29, 229
CERN, 5, 44–46, 127, *128*
  detectors at, 44, 45, 47, *48–49*
  work environment at, 215–216
Chadwick, James, 89
"charge" (other than electric), 39
chargino, 77*t*
Charpak, Georges, 246
Chen, Shao-Long, 237
Chikashige, Yuichi, 161
chirality, 38, 40, 72
  flip, 92
CHOOZ experiment, 238
Chown, Marcus, 225
chronology protection conjecture,
  211
Chung, Daniel, 221, 223
Cicero, 8
Clark, William, 247
Clinton, William Jefferson, 117
closed timelike curve, 204–205,
  206, 211, 222
CMB, 135, *139,* 145, *150,* 153
  and cosmic structure, 148
  discovery of, 131–132
  prediction of, 134
  uniformity of, 137
CMS detector, 47, *48–49*
cobalt, 60, 74
collapse of wave function, 23,
  28–30, 31–32
color charge, 41
Coma Supercluster, 132
compactification, 172, *173*
complementarity, 24
  and ancient Greece, 29

condensate. *See* Higgs condensate
Conrad, Janet, 227
conservation laws, 54
  *See also specific conserved
    quantities*
constructivism, 251, 253
Copenhagen
  Everett visit to, 30
  Heisenberg visit to, 18
Copenhagen (Frayn play), 20
Copenhagen interpretation, 23, 31
cosmic background radiation. *See*
  CMB
cosmic neutrino background, 145,
  241
cosmic rays, 38, 50
cosmological constant, 134, 142
cosmology, 131–153 (Chapter 10)
Cowan, Clyde, 84, *86,* 117, 250
Cowell, Henry, 65
CP symmetry, 239
CP violation, 140, 240
crab nebula, *245*
Csaki, Csaba, 221
cults, 7–16 (Chapter 2), 26, 32
  relation to modern physics, 29
CUORE experiment, 111, 238

Dai, Jin, 173
dark energy, 142–143, 149, 152
dark matter, 79, 80, 141–142, 149,
  186, 231
  detection of, 149
  searches for, 103
Davis, Raymond
  and early neutrino search, 117,
    118
  personality of, 250
  and solar neutrinos, 93, 119–121
DAYA BAY experiment, 239, 241
decoherence, 32, 94
De Crescenzo, Luciano, 26

Demeter (Greek goddess), 8–11, 14
Democritus, 15, 29, 35
Deppisch, Frank, 155, 235, 236
De Rujula, Alvaro, 245
desert (particle), 51, 61, 247
  conferences on, 63–64, 127, 140
Deutsch, David, 228, 229
DeWitt, Bryce, 31
Diamond Head, *2–3,* 4
Dick, Karin, 165
Dicke, Robert, 131, 135
Dienes, Keith, 180–182
di-lepton decay, 231, 235
dimensions, extra, 4, 167–187
  (Chapter 12)
  asymmetric warping of, 221,
    222, 226
  searches for at LHC, 237–238
  warped, 184–187, 219–220
  as wormhole, 214, 220
Dimopoulos, Savas, 77, 170, 177
  See also ADD
Dionysian events, 14–15, 32, 253
Dionysus (Greek god), 11
Dirac, Paul, 24, 83
  in Moscow, 25
Dirac neutrinos, 166
Dirac particles, 91, 92
direct detection
  of cosmic neutrino back-
    ground, 241
  of dark matter, 149
direct production of dark matter, 149
Disney, Walt, 202, 252
Doors (musical group), 12
double beta decay, *105*
Double-CHOOZ experiment, 239,
  241
double seesaw
  See seesaw, inverse
double-slit experiment, *22*
Drees, Manuel, 162

Dreiner, Herbi, 162
duality relations, 174, 234
Dudas, Emilian, 180–182
Dvali, Gia, 170, 177, 186
   See also ADD

E (Eels lead singer), 31
ecstasies, 32
Eddington, Arthur, 118
Eels (musical group), 31
Einstein, Albert, *190*
   and cosmological constant, 134,
      142
   and extra dimensions, 170
   and mass concept, 199
   opposition to quantum
      mechanics by, 21, 29
   at patent office, 188–189
   in Princeton, 202
electromagnetism
   classical, 39, 189
   quantum, 74, 170
   strength of, 169
electron antineutrino, 218
electron volt, 46
Elefsina, Greece, 8
elementary particles. *See* particles
Eleusis, 7–16 (Chapter 2)
Elliott, Steven, 235
Ellis, John, 59
El Monstro detector, 85
energy conservation, 56
   in beta decay, 82–83
energy units, 46*t*
Englert, François, 66, 69
erasure of structures, 148
ergot, 11, 13–14
Erlich, Joshua, 221, 223
ether, 189
Everett, Elizabeth, 30–31
Everett, Hugh, III, 30–32, 228
   and Niels Bohr, 30

Everett, Mark (aka E), 31
EXO-200 experiment, 110, 111, 238

families. *See* particle families
Fardon, Robert, 152
faster-than-light travel, 5, 197–198
Fermi, Enrico, *19,* 250
   and neutrino coinage, 83
   and students of, 83, 86
Fermilab, 5, 127
fermions, 37
Ferrara, Sergio, 77
Feynman, Richard, 20, 30, 118
filaments, 132, 139, 147
fine-tuning problem, 76
Flanz, Marion, 165
flatness, 135
flavor (particle property), 42, 96,
   146, 225
flavor frontier, 236–237
flavor puzzle, 239
force carriers, 40
Frayn, Michael, 20
Freese, Katherine, 221
Frigerio, Michele, 237
Fritzsch, Harald, 58
Fuchs, Klaus, 94
Fukugita, Masataka, 163, 164
fundamental particles. *See* particles

GADZOOKS experiment, 241
Galileo Galilei, 89
GALLEX experiment, 113, 124, 125
gallium, 124
Gamow, George, 134
gauge symmetry, 59
gauge theories, 58
Gehman, Victor, 235
Gell-Mann, Murray, 37, 137, 160
general relativity, 199–201, 205
GENIUS proposal, 111
Georgi, Howard, 58, 59, 77

GERDA experiment, 111, 238
germanium, 125
  isotope 76, 98
Gherghetta, Tony, 180–182
Glashow, Sheldon, x
  and electroweak interaction,
    44, 74
  and neutrinos as probes, 246
  and seesaw mechanism, 160
  and single primordial particle,
    58–59
gluino, 77*t*
gluons, *40*, 41
Gödel, Kurt, 202, *203*, 204–205,
  224, 250
  See also incompleteness
    theorem
Gödel universe, *206*
Goethe, Wolfgang Johann von, 20,
  25
Golden Gate Bridge, 167, *168*
Gol'fand, Yuri, 76
Grand Unified Theories. *See* GUT
grandfather paradox, 209, 228
Gran Sasso Laboratory, 100–101,
  127, *128*
Gratta, Giorgio, 110
gravitation
  in general relativity, 199
  not in Standard Model, 43
  and quantum physics, 175–176,
    180
  as strong force, 177–180
graviton, 170
Greece, ancient, 7–16 (Chapter 2)
Greek philosophy, influence on
  modern physics, 25
  See also individual
    philosophers
Green, Michael, 171
Grojean, Christophe, 221
Grossman, Yuval, 186

ground state, 67, 70
Guralnik, Gerald, 66, 69
GUT, 59–62, 63, 66, 69, 75
  and neutrino mass, 98
  and neutrinos, 91
  neutrinos as probes of, 130
  and predicted proton decay, 126
  scale, 136, 152, 158, 180, 218
  unification of forces in, 78–79
Guth, Alan, 135, 136, 137

Hades (Greek god), 8–9, 11, 14
Hades's grotto, 10
Hadrian, 8
Hagen, Carl R., 66, 69
Hagura, Hiroyuki, 246
Haldeman, Joe, 227–228
Hall, Lawrence, 162
hallucinations, 32
Halzen, Francis, 243
handedness. *See* chirality
Hawai'i, University of, 4
  See also Manoa campus
Hawking, Stephen, 209, 211, 212, 250
Heidegger, Martin, 143
Heidelberg-Moscow experiment,
  102–103
  See also neutrinoless double
    beta decay
Heidelberg University, 97
Heinlein, Robert, 210
Heisenberg, Werner, 17–18, *19*, 20,
  23–25, 250
  at eighteen, 56
  and Greek philosophy, 24–25
  in Helgoland, 20, 24
  and Niels Bohr, 18–19, 23
  and nuclear structure, 89
  and nucleon symmetry, 57, 58,
    83
  threat to his life, 18
  and Wolfgang Pauli, 36, 81–82

helicity, 38, 72
Hellmig, Jochen, 111, 114
Helo Herrera, Juan Carlos, 235
Hemingway, Ernest, 34, 254
Heraclitus, 15
Herman, Robert, 134
Hermann, Armin, 24
Herr Auge detector, 86, 87
Heuer, Rolf-Dieter, 66
Heusser, Gerhard, 111
Hewett, JoAnn, 227
hierarchy problem, 75–76, 169, 185, 231
Higgs, Peter, 65–66
Higgs boson, 50, 66, 76, 80
    exotic decay of, 237
Higgs condensate, 75, 163, 164
Higgs field
    ground state(s) of, 69, 70, 136
    invention of, 66
    and isospin charge, 74
    and neutrino coupling, 158
    and vacuum energy, 143, 157
Higgs mechanism, 66, 67
Higgs triplet, 157
Hilbert, David, 54
Hirsch, Martin, 112, 113, 162, 235, 236
Hitchhiker's Guide to the Galaxy, 33
Hofmann, Albert, 11–15
Hollenberg, Sebastian, 224
Honolulu, 1, *2–3*
Horava, Petr, 173
horizon problem, 139, 221
Hsu, Stephen, 211
Hubble, Edwin, 134, 142
Hubble Telescope, *133*
Huber, Patrick, 246
Huber, Stephan, 186
Hung, P. Q., 149, 152
Huxley, Aldous, 13, 26, 253

Iasion (Greek god), 14
Ibarra, Alejandro, 155
IceCube Neutrino Observatory, 243, 244
IMB detector, 92, 127, 242
incompleteness theorem, 204
indirect detection of dark matter, 149
inflation, 136–137, *138*, 140–141, 153, 241
inflaton (field), 137, 140–141, 143, 147, 166
inhomogeneities in CMB, *138*, *139*, 153
interference, two-slit, 21–22
International Linear Collider, 46
invariance, 77
"Is," 28
isospin, 58, 73, 74, 156

Joliot-Curie, Irène and Frédéric, 93
Jung, C. G., 82
Jünger, Ernst, 11

K2K experiment, 240
    See also T2K experiment
Kachelriess, Michael, 161, 215
Kajita, Takaaki, 115
Kaluza, Theodor, 170, 171
Kaluza-Klein excitations, 186, 226, 237
Kamiokande detector, 92
    See also Super-Kamiokande
KamLAND, 129–130
KamLAND-Zen experiment, 111
KATRIN
    detector, 107, *108–109*
    experiment, 107, 238
KEK laboratory, 127
Kellogg, J. M., 85
Kephart, Thomas, 214, 215
Kepler, Johannes, 244

Kerouac, Jack, 249
Kibble, Tom, 69
Kirsten, Till, 113, 124
Klapdor-Kleingrothaus, Hans
  Volker
  and "desert" conferences, 63, 127
  and neutrinoless double beta
    decay, 98, 103, 110
  personality of, 250
  and theoretical studies, 111,
    113, 163
Klein, Felix, 54
Klein, Oskar, 170, 171
KM3NET telescope, 244
Kolb, Edward W., 140
Kom, Steve, 235
Kovalenko, Sergey, 112, 113, 235
Kuhn, Thomas S., 24, 251, 252
Kuzmin, Vadim, 121, 164
kykeon, 10, 11, 14

Large Hadron Collider. *See* LHC
Learned, John, 126, 127, 242, 243,
  246
Leary, Timothy, 13
Lederman, Leon, 92
Lee, T. D., 71
Leigh, Robert, 173
Lemaitre, Georges, 134, 142, 205
leptogenesis, 160, 165
lepton number, 91, 112
  conservation of, 91
  violation of, 182
leptons, 37, 74
leptoquark forces, 158
leptoquarks, 62
Leser, Philipp, 237
Lewis, Meriwether, 247
LHC, 46, *47,* 231, *232–233,* 234
LHCb detector, 47
light cones, 195–196, *197*
like sign di-lepton signal, 231, 235

Likhtman, Evgenyi, 76
Linde, Andrei, 136, 167
Lindner, Manfred, 163, 165
linear collider. *See* International
  Linear Collider; SLAC
Lir (Celtic god), 65
local group of galaxies, 132
Lono (Hawaiian god), 4
Los Alamos lab
  and LSND, 114, *217,* 227, 235
  and neutrino detection, 84, 85,
    87
Louis, William, 227
LSD, 11–14
LSND experiment, 114, 217, 218,
  219, 227, 240
lysergic acid, 11–14

M-theory, 174
Ma, Ernest, 237
Mach, Ernst, 189, 199
MACRO experiment, 102, 136
Magellanic Clouds, 132
magic mushrooms, 12, 14
magnetic monopoles, 137
magnetism, 68–69
Majorana, Ettore, 87–91, 156, 250
  disappearance of, 89–91
Majorana particles, 91, 92, 104,
  118, 238
  and double beta decay, 112,
    234–235
  and neutrino fluctuations, 163
majoron, *159,* 161
Majorovits, Bela, 111
Major Tom, 248–250
Manoa campus, 4, 220
"many worlds," 31–32, 33, 253
March-Russell, John, 180–182
mass
  gravitational and inertial, 199
  origin of, 67

mass hierarchy, 239
Max Planck Institute for Nuclear
    Physics. *See* MPIK
McDonald, Arthur, 130
MEG experiment, 236
mesons, 41, 125
Mexican hat potential, 67
Michelson, Albert Abraham, 189
Micu, Octavian, 224
Mikheev, Stanislav, 122, 123
Milky Way, 132
MiniBooNE experiment, 226, 227,
    240
Minkowski, Peter, 58, 59, 160
MINOS experiment, 127
mirror symmetry, 52, *53*, 72
mixing of states, 94–95
Mohapatra, Rabi, 112, 160, 161,
    162, 163
momentum conservation, 55–56
Morley, Edward, 189
Morris, Michael, 208
Moses, 51
MPIK, 113, 124
MSW effect, 123, 125, 129, 215, 219,
    225
muon, 37, 44, 74, 77*t*, 101, 102
muon antineutrino, 218
muon neutrino, 37*t*, 94, 128
    atmospheric, 126
    discovery of, 92
    and oscillation, 95–96, 117, 123,
        129
Murayama, Hitoshi, 117, 166, 167
Mussolini, Benito, 90

negative energy, 185, *208*, 211,
    225
Nelson, Ann, 152
NEMO experiment, 244
NESTOR experiment, 244
Neubert, Matthias, 186

neutralino, 77*t*, 79
neutrino astrophysics, 242
neutrino coinage, 83
neutrino-electron doublet, 59
neutrinoless double beta decay, 105
    evidence for, 110
    and Gran Sasso experiment,
        98–99, 102–103, 111
    and handedness, 160
    implications of, 113, 231
    and nature of neutrinos, 234–235
    and other physics, 145, 161, 162
    and possible new physics, 112,
        182
neutrino oscillations, 216–217
    See also neutrinos: mixing of
neutrinos, 81–96 (Chapter 7),
    154–166 (Chapter 11)
    atmospheric, 126, 216
    Dirac, 166
    families of, 94
    flavors of, 117, 146, 147
    masses of, 71, 130, 146, 148
    mass generation of, 154–155,
        159, 162, 234, 236, 237
    mixing of, 130, 238
    as probes, 242
    right-handed, 157, 158, 165, 186
    small mass of, 180–181
    from supernovas, 102
    See also atmospheric neutrinos;
        cosmic neutrino background;
        neutrino oscillations; solar
        neutrinos; sterile neutrinos
neutron, 41
    decay of, *95*, 103–105, 112, 118
    in early universe, 147
    as nuclear constituent, 35, 89
    production of, 84, *85*, 86, 241
    as proton's other face, 57–58,
        59
Newton, Isaac, 54, 89

Nietzsche, Friedrich, 7, 15, 16, 252, 253
Nobel laureates, x, 44, 58, 74, 83, 86, 144, 164, 246
Noether, Emmy, 52, 54–56, 58
NOVA experiment, 240
nucleosynthesis, 146
nucleus, atomic, 84, *95*
    and atomic structure, 35–37
    structure of, 41, 89

Occam's Razor, 231
Olympus, Mount, 8–9
"One," 26–29, 229
OPERA experiment, 129, 227
ORCA experiment, 244
oscillation, 216–217
    See also neutrinos: mixing of

Pakvasa, Sandip, 122, 126, 130, 216, 220, 221, 246
Pal, Palash, 121, 162
Palomares-Ruiz, Sergio, 213
Panella, Orlando, 112
Panigrahi, Prasanta, 246
paradigm, shift of, 24, 251, 252
paradoxes, 5, 209, 210, 228
parallel universes, 30–32, 229
Parmenides, 15, 25–29, 229, 253
Parmenides (Plato dialogue), 25
particle families, 37*t*, 38, 39, 92, 146
particle loop, 162
particles, 34–50 (Chapter 4)
    left-handed, 73
    massless, 72
    right-handed, 73
Päs, Frigga, drawing by, *190*
Paschos, Emmanuel A., 165
Pauli, Wolfgang, *19,* 36, 73, 74, 176, 250
    and neutrino postulate, 81–83
Pauli principle, 36–37

Pausanias, 8
Peccei, Roberto, 161
Peierls, Rudolph, 71, 73
Penzias, Arno, 131, 132, 135
Persephone (Greek goddess), 8–11, 26
Petcov, Sergey, 155
Phillips, Roger, 122
photino, 77*t*
photon, 77*t,* 145
    as decay product, 236
    as field quantum, 40, 41
    as massless particle, 42, 72
Pidt, Daniel, 236
Piepke, Andreas, 110
Pilaftsis, Apostolos, 165, 182
Pindar, 8
PINGU experiment, 244
pion
    decay of, 240
    quark composition of, 41, 125–126
Planck mass, 169
PLANCK satellite, *150–151,* 153, 241
Planck scale, 135
plasma, 69, 134, 141, 220
Plato, 8, 15, 249
    his allegory of the cave, 15, 69–70
    influence on Heisenberg, 24
Plümacher, Michael, 165
Polchinski, Joseph, 173, 174
Pontecorvo, Bruno, *88,* 93–94, 117, 118, 120
Popper, Karl, 176, 250, 251
Porod, Werner, 162, 236
positron, 39, 84, 91
    production of, *85,* 118
Powers, Thomas, 18
probability in quantum mechanics, 21–23, 29
Project Poltergeist, 86, *87*

proton, 41
    acceleration of, 45–47
    in early universe, 147
    fusion reactions of, 120, 121
    as neutron's other face, 57–59, 83
    as nuclear constituent, 35, 89
    possible decay of, 62, 63, 126
    production of, 95, 103–105
    transformation of, 84, 85, 118
proton decay, search for, 62–63, 126
psychedelic drugs, 15, 32
    See also ergot; LSD; Salvia
        divinorum

quanta of fields, 40
quantum field theory, 70
quantum fluctuations, 42, 225,
    234, 235
quantum numbers, 36, 56
quantum physics, 17–33 (Chapter 3)
    and gravity, 175–176, 180
    neo-Platonic nature of, 32
quarks, 37
    mixing of, 123, 237
    and symmetry, 56

R-parity, 79, 112, 162, 235, 236
radioactivity, 40, 95
    background, 99, 102–103,
        110–111
    and neutrino search, 125
    of specific isotopes, 99, 113
Ramond, Pierre, 76, 160
Randall, Lisa, 183–186, 221
Randall-Sundrum models,
    184–186, 221
Ratz, Michael, 165
Redelbach, Andreas, 156, 236
red shift, 134, 142
Reines, Frederick, 84, 85, 86, 117,
    127, 250
RENO experiment, 239, 241

Republic (Plato dialogue), 70
Rodejohann, Werner, 155
rotational symmetry, 52, 53
Rubakov, Valery, 164, 165
Rubbia, Carlo, 44–45, 250
Ruck, Carl, 13
Rückl, Reinhold, 156
Russell, Bertrand, 34

Sacharov, Andrei Dmitrievich, 144
Sagan, Carl, 207
SAGE experiment, 124
Salam, Abdus, 44, 70–71, 72, 73, 74
Salvia divinorum, 13
Sanami, Toshiya, 246
Santamaria, Arcadi, 162
Sarkar, Utpal, 165, 246
scattering concept, 68
Schechter, Joel, 112, 160
Schechter-Valle theorem, 234
Scherk, Joel, 170
Schilling, Peter, 248
Schmidt, Ivan, 112
Schrödinger, Erwin, 23, 24, 25
Schrödinger's cat, 23, 27, 91
Schwartz, Melvin, 92
Schwarz, John, 170
Schwarzschild radius, 200–201
Sciascia, Leonardo, 89, 90
seesaw, inverse, 161, 234
seesaw-II, 157
seesaw mechanism
    alternatives to, 161
    double, 161, 234
    in extra dimensions, 182
    and neutrino mass, 152,
        155–157, 163
    and SUSY, 235
Senjanovic, Goran, 160
Shafi, Qaisar, 186
Shaposhnikov, Mikhail, 164
simultaneity, relativity of, 194, 229

Singh, Anupam, 152
Sisyphus, 252
SLAC, 168
Slansky, Richard, 160
Smirnov, Alexei, 113, 122, 123, 165, 186, 215
sneutrino, 77t, 112, 166, 235, 241
SNO, 129
Socrates, 26, 249, 250, 252, 253
solar neutrinos, 113, 218
    and Davis experiment, 119
    and extra dimensions, 216
    oscillation of, 114
    within the sun, 122–124, 219
Sommerfeld, Arnold, 81
Song, Liguo, 214
Sophocles, 8
space-time
    diagrams, 195
    warping of, 201
spectrometer, 106, 107, *108–109,* 189
speed of light
    constancy of, 192
    as speed limit, 197
sphaleron, 165, 166
spin, 36
squark, 77t, *173*
Stalin, Joseph, 54
Stancu, Ion, 227
Standard Model, 38, 44, 45
    completion of, 80
    extensions of, 160
    failure of, 117, 119
    and mass, 67, 70
    and neutrino flavor, 146
    and neutrinos, 72
    no mirror symmetry in, 72
    physics beyond, 130
    possible limitations of, 45
    and spontaneous symmetry
        breaking, 75, 80
    SUSY counterparts of, 77t

Stanford Linear Accelerator
    Center, 168
Star Trek, 204, 209
Steinberger, Jack, 92
Steinhardt, Paul, 136
sterile neutrinos, 157, 161
    evidence for, 240
    and extra dimensions, 216, 219, 220, 226
    great mass of, 218
    and time travel, 224, 225
Stockum, Willem Jacob van, 205
Strecker, Herbert, 99
string scale, 158
string theory, 170–171, 174–179
Sudbury Neutrino Observatory
    (SNO), 129
Sugawara, Hirotaka, 246
Sundrum, Raman, 184–186, 221
Super-Kamiokande, *116,* 117, 127, 240, 241, 242
superluminal speed, 5, 197–198
Supernova 1987A, 92–93, 242
superposition, 21–22
superstrings, 170
superstring theory. *See* string theory
supersymmetry. *See* SUSY
surfing, 1–6 (Chapter 1)
Susskind, Leonard, 167, 176, 223
SUSY, 63–64, 76–80
    frontier, 235–236
    names of counterpart particles, 77t
    neutrino partners, 241
    and neutrinoless double beta
        decay, 112–113
    particle masses, 78
    partners, 182
    and R-parity violation, 162
    scale, 152, 234
Sutton, Christine, 249
Suzuki, Hiroshi, 166

Suzuki, Mahiko, 162
Swarup, Vikas, ix
symmetries, 51–64 (Chapter 5)
    See also specific symmetries
symmetry breaking, 63, 65–80
    (Chapter 6)
    spontaneous, 66–67, 69
synthesia, 12

T2K experiment, 239–240
Takita, Masato, 127
tau, 37, 74, 77t
tau neutrino, 94, 96, 117, 123, 126,
    128, 129
Taubes, Gary, 44
Telesterion, 10, 13
t'Hooft, Gerard, 164
Thorne, Kip, 30, 207, 208
Timaeus (Plato dialogue), 16, 17,
    29, 56
time
    dilation, 192, *193*
    nature of, 188–201 (Chapter
        13), 228
    translation, 56
time travel, 5, *198,* 202–212
    (Chapter 14), 213–229 (Chapter
    15)
Tipler, Frank J., 205–206, 224
Tomas, Ricard, 161
translational symmetry, 55–56
tritium
    beta decay of, 106
    and neutrino mass, 106
two-slit interference, 21–22
Tye, Henry, 136

UA1 experiment, 44, 215
UA2 experiment, 44, 45, 215
uncertainty principle, 42
    energy-time version of, 43,
        155

underground detectors, 97–114
    (Chapter 8)
unification
    of forces, *62*
    of particles, 51–64 (Chapter 5)
    and Plato, 57, 58
universe
    accelerated expansion of, 142
    expansion of, 134
    as a whole, 28
Uranverein (German atomic
    project), 18, 83

vacuum
    energy of, 152
    fluctuations, 79
    polarization of, 60, *61*
    in quantum field theory, 70
Vagins, Mark R., 241
Valle, Jose, 112, 160, 161, 162, 163,
    236
Van der Meer, Simon, 44–45
van Gogh, Vincent, 253–254
visions, 32
Visser, Matt, 221, 224
Volkov, Dmitri, 76

W bosons, *40,* 41–42
    discovery of, 44
    and energy-time uncertainty,
        43
    relation to Higgs mass, 75
    and SNO, 129
Waikiki, 1, *2–3,* 223
warping. *See* dimensions, extra;
    space-time
Wasson, R. Gordon, 13
wave-particle duality, 21–22
Weiler, Thomas, 212
    and cosmic neutrino back-
        ground, 145
    and double beta decay, 113

and extra dimensions, 215, 216,
221
and neutrino mass, 214
and neutrino oscillation, 126, 130
Weinberg, Steven, 44, 74, 163
Weiner, Neal, 152
Weizsäcker, Carl Friedrich von,
25–29
Wess, Julius, 77
Wheeler, John Archibald, 30, 31
Whisnant, Kerry, 122
White, Edward, 248
Wiesenfeldt, Sören, 234
William of Occam, 231
Wilson, Robert, 131, 132, 135
Wilson, Robert Rathbun, 245
Witten, Edward, 174
Woit, Peter, 176
Wolfenstein, Lincoln, 121–122,
123, 250
wormhole, 207–209, 214
Wright, David, 165
Wu, Chien Shiung, 73–74

Yanagida, Tsutomu, 160, 163, 164,
166
Yang, C. N., 71
Yokoyama, Jun'ichi, 166
Yurtsever, Ulvi, 209

Z boson, 40, 41–42, 162
and cosmic neutrino back-
ground, 145
decay of, 218
and energy-time uncertainty,
43
related to HIggs mass, 75
and SNO, 129
Zaphod Beeblebrox, 33
Zaslavskii, Oleg, 91
Z-burst, 145, 241
Zee, Anthony, 162, 246
Zeus (Greek god), 8–9
Zichichi, Antonino, 100, 250
zinos, 162
Zumino, Bruno, 77
Zweig, George, 37